The Hoek-Brown Failure criterion—From theory to application

Jianping Zuo · Jiayi Shen

The Hoek-Brown Failure criterion—From theory to application

Springer

Jianping Zuo
School of Mechanics and Civil Engineering
China University of Mining & Technology-Beijing
Beijing, China

Jiayi Shen
Institute of Port, Coastal and Offshore Engineering
Zhejiang University
Hangzhou, China

ISBN 978-981-15-1771-6 ISBN 978-981-15-1769-3 (eBook)
https://doi.org/10.1007/978-981-15-1769-3

This Springer imprint is published by the registered company Springer Nature Singapore Pte Ltd.
The registered company address is: 152 Beach Road, #21-01/04 Gateway East, Singapore 189721, Singapore

Path is shown up only when thousands of people walk through.
Sir Lu Xun

To practitioners of rock mechanics and mining engineering.

Preface

The Hoek-Brown (HB) failure criterion is one of the most broadly adopted failure criteria to estimate rock mass strength in rock mechanics and mining engineering. Over the past 40 years, the HB criterion has been applied successfully to a wide range of intact and fractured rock types.

The book aims to help researchers, engineers and research students who work in the area of rock mechanics and mining engineering. By reading this book, readers can get to know the theory and principle of the HB failure criterion, methods or guidelines for estimating the HB input parameters, and the methodology of application of the HB criterion in rock engineering projects.

The historical development of the HB failure criterion and associated rock mass properties are presented in Chap. 1. The state of the art of the theoretical deviation of the HB criterion from micro-mechanics principles is introduced in the Chaps. 2–4. It is proved that the constant m_i in the HB has physical meaning and is able to relate the microscopic damaged crack characteristics to macro-failures successfully. After that, methods and guidelines for estimating the HB input parameters (the unconfined compressive strength σ_{ci}, the constant m_i, Geological strength index GSI and blasting damage factor D) are introduced in Chaps. 5–7. Finally, five cases about the implementation of the HB criterion into coal mining engineering, rock engineering structure and rock slope stability analysis are presented in Chaps. 8–12.

The book is written in an easily readable style. Academics can quickly obtain an overview of the state of the art of the theory of the HB criterion by reading the book before they advance their researches on the topics related to rock failure criteria. Geotechnical engineers can select appropriate HB input parameters for the design and analysis of rock engineering projects with the help of the principles introduced in this book. Research students may use the book as a textbook to learn the principle of rock mechanics related to rock mass properties.

Beijing, China — Jianping Zuo
Hangzhou, China — Jiayi Shen

Acknowledgements

The authors would like to thank the following organizations for their supports:

The state key Laboratory of Coal Resource and Safe Mining, China University of Mining and Technology-Beijing.
Brown University.
Colorado School of Mines.
Lawrence Berkeley National Laboratory.

This book was supported by Beijing Outstanding Young Scientist Program (BJJWZYJH01201911413037); the National Natural Science Foundation of China (grant No. 51622404, 11572343, 51374215, 51504218), Yueqi outstanding scholar Award Program by CUMTB, Outstanding Young Talents of "Ten Thousand People Plan" (grant No. W02070044), National Basic Research Program of China (973 Program) (grant No. 2010CB732002), Beijing Major Scientific and Technological Achievements into Ground Cultivation Project (grant No. Z151100002815004), National Excellent Doctoral Dissertation of China (201030), China Postdoctoral Science Foundation (201030), State Key Research Development Program of China (grant No. 2016YFC0801404), Fok Ying Tung Education Foundation (142018) and Coal Mines Corporations. The State Key Laboratory for GeoMechanics and Deep Underground Engineering, China University of Mining and Technology (SKLGDUEK1808).

The authors would also like to thank the supports and encouragement from the following individuals:

The first author would like to thank Academician Heping Xie, Academician Suping Peng, Prof. Hongwei Zhou, Prof. Yang Ju for the lead and guide in the area of rock mechanics and rock engineering. The author appreciates the help from Dr. Hongtao Li and Dr. Huihai Liu. The authors also thank the support and help from the following students: Yunqian Jiang, Xu Wei, Shunyin Hu, Yunjiang Sun, Jintao Wang, Jinhao Wen, Yubo Zhou, Yue Shi, Hongqiang Song, Zijie Hong, Zhengdai Li, Meilu Yu, Changning Mi et al.

The second author would like to thank Dr. Murat Karakus and Dr. Chaoshui Xu, (University of Adelaide, Australia) for their contribution to the contents on the determination of rock properties and the analyses of slope stability, which become parts of this book in Chaps. 5, 10 and 11. Thus, many thanks go to Murat and Chaoshui for their contribution.

Contents

1 The Hoek-Brown Failure Criterion 1

1.1 Introduction 1

1.2 The Hoek-Brown Criterion for Intact Rocks 2

1.3 The Generalized Hoek-Brown Criterion 5

1.4 Strength and Deformation of Rock Masses Based on the HB Criterion 6

1.5 Mohr-Coulomb Parameters Based on the HB Criterion 8

1.5.1 Instantaneous Shear Strength Parameters 9

1.5.2 Equivalent Shear Strength Parameters 12

1.6 The Application of the HB Criterion 14

References 15

2 A 2-D Theoretical Derivation of the Hoek-Brown Criterion 17

2.1 Introduction 17

2.2 Fundamental Hypothesis and 2-D Model 18

2.3 Critical Crack Propagation Analysis 19

2.4 Micro-failure Orientation Angle α of Rock 21

2.5 Rock Failure Characteristic Parameter 23

2.6 A Theoretical Derivation of the Hoek-Brown Failure Criterion 25

References 27

3 A 3-D Theoretical Derivation of the Hoek-Brown Criterion 29

3.1 Introduction 29

3.2 Fundamental 3-D Model for Cracked Brittle Materials 30

3.3 Propagation Condition of Disk-Shaped Crack 31

3.4 The Attitude Angle Parameter φ to Character Micro-fracturing Orientation (Zuo et al. 2015) 32

3.5 The Failure Characteristic Parameter for Brittle Materials 34

3.6 A Theoretical Derivation of the Hoek-Brown Failure Criterion 36

3.7 The Relationship Between Other Trigonometric Functions and Stress . . . 39
3.7.1 Relationship Between sinα and Stress . . . 40
3.7.2 Relationship Between tanα and Stress . . . 42
3.7.3 Relationship Between cotα and Stress . . . 44
References . . . 46

4 The Hoek-Brown Brittle-Ductile Transition Analysis . . . 49
4.1 Introduction . . . 49
4.2 Experimental Researches on the Brittle-Ductile Transition of Rock . . . 50
4.3 Microstructure Effects of Rock Minerals . . . 53
4.4 Brittle-Ductile Transition Analysis Based on H-B Criterion . . . 58
References . . . 61

5 The Hoek-Brown Constant m_i . . . 63
5.1 Introduction . . . 63
5.2 Indicators for the Evaluation of Existing Methods . . . 65
5.3 Existing Methods for Estimating m_i Values . . . 66
5.3.1 Regression Method . . . 66
5.3.2 Guidelines . . . 69
5.3.3 R Index . . . 69
5.3.4 UCS Based Model . . . 72
5.3.5 TS Based Model . . . 73
5.3.6 Methods Based on Other Rock Properties . . . 77
5.4 Comparison of Prediction Performance of Various Models . . . 78
References . . . 82

6 The Geological Strength Index . . . 85
6.1 Introduction . . . 85
6.2 Background of GSI . . . 86
6.3 The General GSI Chart . . . 88
6.4 Quantification of GSI Charts . . . 90
6.4.1 GSI Charts by Sonmez and Ulusay (1999) . . . 91
6.4.2 GSI Charts by Cai et al. (2004) . . . 93
6.4.3 GSI Charts by Hoek et al. (2013) . . . 97
6.5 When Not to Use GSI . . . 101
References . . . 103

7 The Blast Damage Factor *D* . . . 105
7.1 Introduction . . . 105
7.2 Guidelines for Estimating Disturbance Factor *D* . . . 106
7.3 Guidelines for Estimating Disturbance Zone Thickness *T* . . . 108
7.4 Effect of *D* on Rock Mass Properties . . . 109
References . . . 115

8 A Gas-Mechanical Coupled Hoek-Brown Criterion 117
8.1 Introduction 117
8.2 Typical Stress-Strain Relationship for Fractured Coal 118
8.3 A Model for Fractured Coal Including Gas Based on Hoek-Brown Criterion 121
8.4 Microscopic Observations on the Crack Distribution of Coal 125
8.5 Constitutive Equation for Coal Containing Pressure Gas 127
8.6 Sensitive Analyses and Comparison with Experimental Data . . . 130
References 134

9 An UCS Model for Anisotropic Blocky Rock Masses Satisfying the Hoek-Brown Criterion 137
9.1 Introduction 137
9.2 Methods for Estimating Rock Mass Strength Using Laboratory-Size Specimens 139
9.2.1 Rock Specimen Configuration 139
9.2.2 Quantifying the Degree of Fracturing Using Equivalent GSI 140
9.3 Calibration of UDEC Rock Mass Models 141
9.3.1 Calibration for Intact Rock Model 142
9.3.2 Calibration for Rock Mass Models 143
9.4 Rock Mass Configurations for Numerical Simulations 147
9.5 Numerical Simulation Results 147
9.6 Modified Anisotropic UCS Model for Jointed Rock Masses . . . 152
9.6.1 Comparison of UCS Obtained from Empirical and Numerical Models 152
9.6.2 Anisotropic Weighting Factor 153
9.6.3 The Modified UCS Model 155
9.7 Conclusions 158
References 158

10 Non-linear Shear Strength Reduction Method for Slope Stability Based on the HB Criterion 161
10.1 Introduction 161
10.2 Instantaneous Shear Strength of the HB Criterion 163
10.3 Non-linear SSR Method for the HB Criterion 165
10.4 Convergence Criterion in the 3D Models 167
10.5 Boundary Conditions in 3D Models 170
References 175

11 Chart-Based Slope Stability Assessment Using the Generalized Hoek–Brown Criterion 177
11.1 Introduction 177
11.2 Review of Existing Rock Slope Stability Charts Based on the HB Criterion 179
11.3 Proposed Stability Charts for Rock Mass Slopes 182
11.3.1 Theoretical Relationship Between SR and FOS 183
11.3.2 Slope Stability Charts Based on Slope Angle $\beta = 45°$ 186
11.3.3 The Disturbance Weighting Factor f_D 189
11.3.4 The Slope Angle Weighting Factor f_β 192
11.3.5 The Use of the Proposed Stability Charts 195
11.4 Slope Cases Application 196
11.5 Conclusions 197
References 198

12 The Effects of Blast Damage Zone Thickness on Rock Slope Stability 201
12.1 Introduction 201
12.2 Parallel Layer Model (PLM) 205
12.3 Comprehensive Stability Analysis Based on PLM 206
12.4 The Blast Damage Zone Thickness Weighting Factor f_T 206
12.5 Stability Model Based on f_T and Existing Stability Charts 214
12.6 Slope Cases Application 219
12.7 Conclusions 220
References 221

Appendix 223

About the Authors

Jianping Zuo obtained his Ph.D. from China University of Mining and Technology (Beijing), China in 2006. He is a full Professor of Engineering Mechanics and the Dean of School of Mechanics and Civil Engineering in CUMTB. He has visited Brown University as a senior visiting scholar and Lawrence Berkeley National Laboratory as a visiting scholar. He has been involved in rock mechanics and mining engineering research, consulting and teaching for more than 13 years. He is in charge of and/or participates in 12 scientific research projects, funded by National Natural Science Foundation of China, National Basic Research Program of China (973 Program), Beijing Major Scientific and Technological Achievements into Ground Cultivation Project, the 111 Project and Coal Mines Corporations. He is the author or co-author of more than 130 peer review journal papers and 20 conference papers. He has applied for 28 patents, and received 12 Natural Science and Technology Progress Awards.

Jiayi Shen received his Ph.D. in rock mechanics at The University of Adelaide (U of A) in Australia. After his Ph.D., he worked as a research associate in mining rock mechanics in U of A for a short period. Then, he was employed as a Lecturer of rock mechanics in the Zhejiang University in 2013 and then Associate Professor in 2015. Dr. Shen has more than 10 years of experience in rock mechanics for mining and civil engineering. His major interests are stability of rock masses and failure mechanics of rock masses. He has more than 20 scientific publications to his name. Dr. Shen has practical expertise in stability rock slopes and underground excavations, in situ measurements and interpretations.

Chapter 1
The Hoek-Brown Failure Criterion

Abstract One of the most challenging problems encountered in the design of rock engineering projects is the estimation of strength and deformation properties of rock masses containing joints, bedding planes, and other geotechnical defects. Currently, one of the commonly used empirical failure criteria in rock engineering is the Hoek-Brown (HB) failure criterion. The main breakthrough of the HB criterion is that it opened up for empirically estimating rock mass strength and deformation properties at the engineering scale. The HB criterion started with intact rock properties and then it was extended to estimate the rock mass strength by using the Geological Strength Index (GSI) and the disturbance factor D to reduce intact rock properties. A number of empirical models have been proposed in parallel with the HB failure criterion to estimate the rock masses properties such as deformation modulus, UCS and shear strength parameters.

1.1 Introduction

The Hoek-Brown (HB) failure criterion is an empirical relation that characterizes the stress conditions that lead to failure in intact rocks and rock masses. The main breakthrough of the HB criterion is that it opened up for empirically estimating rock mass strength at the engineering scale. The HB criterion started with intact rock properties and then it was extended to estimate the rock mass strength by using the Geological Strength Index (GSI) and the disturbance factor D to reduce intact rock properties. The fact that the HB criterion has been extensively applied with reasonable success in a variety of engineering projects, such as slope, tunneling and foundation, all over the world would indicate that this failure criterion is feasible enough to provide appropriate designs in engineering practice. Therefore, a number of correlations have been proposed in parallel with this failure criterion, which can be used to estimate other relevant rock mass parameters, such as deformation modulus, uniaxial compressive strength and shear strength parameters of rock masses. This chapter provides an introduction of the HB criterion corresponding to empirical relations which widely used for estimating the mechanical properties of rock masses.

J. Zuo and J. Shen, *The Hoek-Brown Failure criterion—From theory to application*,
https://doi.org/10.1007/978-981-15-1769-3_1

1.2 The Hoek-Brown Criterion for Intact Rocks

It is known that the failure in brittle materials, such as concretes and rocks, always generate from micro-cracks in the intact materials. In rocks, these micro-cracks or flaws are usually inter-granular cracks or grain boundaries and tensile cracks that propagate from their tips when frictional sliding occurs along the flaws (Hoek and Brown 2018).

Griffith (1921) noted that tensile failure in brittle materials, such as glass, initiates at the tips of defects that can be represented as flat elliptical cracks. Griffith's original work dealt with a fracture in materials subjected to tensile stresses, but later Griffith (1924) proposed a non-linear compressive failure criterion for brittle materials using biaxial compression tests.

Based on the Griffith failure criterion, Hoek and Brown (1980a, b) proposed an empirical equation (see Eq. 1.1) to fit a wide range of triaxial tests on intact rocks:

$$\sigma_1 = \sigma_3 + \sigma_{\mathrm{ci}}\left(m_{\mathrm{i}}\frac{\sigma_3}{\sigma_{\mathrm{ci}}} + 1\right)^{0.5} \tag{1.1}$$

The equation contains two intact rock properties, namely, the uniaxial compressive strength (UCS) of the intact rock σ_{ci} and the HB constant m_{i}. Values of m_{i} depend on many factors, such as grain sizes and mineral compositions. In order to get the best fit of the failure envelope of a rock, the values of m_{i} should be calculated from triaxial data using regression analyses. This involves laboratory tests on rock specimens which are carefully collected and prepared. Generally, the core should be ensured to be gained from homogeneous rocks. Triaxial tests should be carried out based on the the ISRM suggested methods (Ulusay and Hudson 2007).

In general, m_{i} is considered as a curve-fitting parameter for getting the HB failure envelope. However, researches by Zuo et al. (2008) and Zuo et al. (2015) showed that m_{i} is not just a curve fitting parameter, but has physical meanings and can be derived from micro-mechanics principles. The referred m_{i} model is expressed in Eq. (1.2).

$$m_i = \frac{\mu\sigma_{ci}}{\beta\sigma_t} \tag{1.2}$$

where σ_{ci} and σ_{t} is the UCS and tensile strength (TS) of intact rocks, respectively, μ is the coefficient of friction for the pre-existing sliding crack surfaces, and β is an intermediate fracture mechanics parameter that can be obtained from experimental data. The state of the art of the theoretical derivation of the HB criterion from micro-mechanics principles will be introduced in the Chaps. 2–4. Various methods for calculating m_{i} will be introduced in Chap. 5.

It should be noted that the HB criterion is proposed to deal with shear failure in rocks. Therefore, the HB criterion is only applicable for confining stresses within

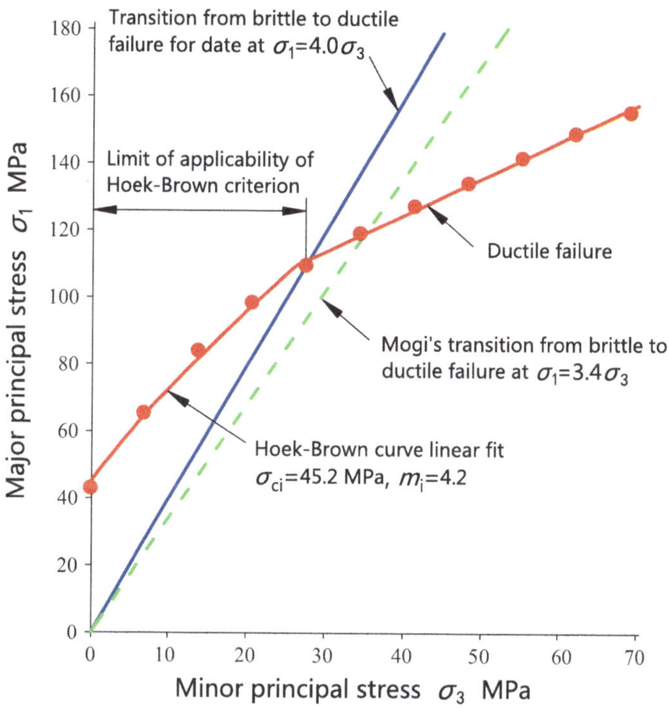

Fig. 1.1 Limit of applicability of the HB criterion (after Hoek and Brown 2018)

the range defined by $\sigma_3 = 0$ and the transition from shear to ductile failure, as shown in Fig. 1.1. Triaxial test data of Indiana limestone by Schwartz (1964) in Fig. 1.1 shows that the applicability of the HB criterion is determined by the transition from shear to ductile failure at approximately $\sigma_{ci} = 4.0\ \sigma_3$. Mogi (1966) found that the average transition is defined as $\sigma_{ci} = 3.4\ \sigma_3$, which is a convenient guide for the selection of the maximum confining pressure for triaxial tests of intact rocks.

Figure 1.1 shows that tensile failure ($\sigma_3 < 0$) is not included by the HB criterion. However, tensile failure is quite common in some rock engineering problems. Hoek and Martin (2014) suggested that an HB failure envelope with a tensile cut-off, which is based on the generalized Griffith failure criterion theory proposed by Fairhurst (1964), can provide an effective solution for practical rock engineering purposes, as illustrated in Fig. 1.2.

It should be noted that the tensile strength of intact rock should be tested from direct tensile tests. However, direct tensile tests are not routinely carried out as a standard procedure in many rock testing laboratories because of the difficulty in the specimen preparation. Indirect methods, on the other hand, such as Brazilian tests are widely used to estimate the tensile strength. Read and Richards (2014) found that direct tensile strength (DTS) values may be taken as 0.9 times of the Brazilian

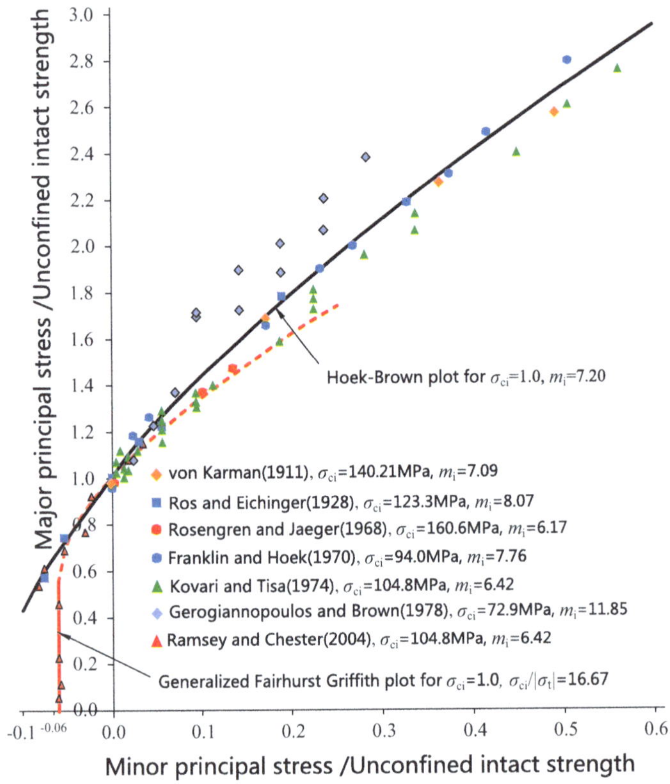

Fig. 1.2 Tensile failure and the HB criterion for shear failure (Hoek and Brown 2018)

tensile strength (BTS) values. In the absence of tensile strength, the following relations can be used for primary estimation, which is suggested by Hoek and Brown (2018), as shown in Eq. (1.3).

$$|\sigma_t| = \frac{\sigma_{ci}}{0.81m_i + 7} \tag{1.3}$$

The HB criterion given by Eq. (1.1) is expressed in terms of the major and minor principal stresses. The intermediate principal stress σ_2 is equal to σ_3 in the triaxial compression test and do not consider effect of the intermediate principal stress σ_2 on the rock strength. However, it is known that when σ_2 is not equal to σ_3, increasing the values of σ_2 will produce small but measurable increases in the peak strengths of intact rock samples tested in true triaxial compression tests. However, Hoek and Brown (1980a) noted it was reasonable to ignore the influence of σ_2 on rock material strength to keep the criterion as simple as possible.

1.3 The Generalized Hoek-Brown Criterion

The generalized Hoek-Brown criterion (Hoek et al. 2002) for the estimation of rock mass strength is expressed as follows:

$$\sigma_1 = \sigma_3 + \sigma_{ci}\left(m_b \frac{\sigma_3}{\sigma_{ci}} + s\right)^a \tag{1.4}$$

where, m_b, s and a are the input parameters which can be estimated from the Geological Strength Index (GSI), disturbance factor D and intact rock constant m_i, as follows:

$$m_b = m_i \exp\left(\frac{GSI - 100}{28 - 14D}\right) \tag{1.5}$$

$$s = \exp\left(\frac{GSI - 100}{9 - 3D}\right) \tag{1.6}$$

$$a = \frac{1}{2} + \frac{1}{6}\left(e^{-GSI/15} - e^{-20/3}\right) \tag{1.7}$$

The generalized Hoek-Brown criterion was proposed to estimate the strength of rock masses which comprised of interlocking angular blocks, and assume the failure of rock mass is governed by intact rock sliding and rotation without many intact rock failures under the state of low or moderate confining stresses.

The GSI classification contains two principal factors, namely, the rock structure and the surface condition of the discontinuities which have significant influences on the mechanical properties of rock masses. Guidelines for the selection of GSI will be discussed in Chap. 6.

When slopes and tunnels are excavated in rock masses, removal of rock masses by blasting or mechanical excavation results in stress relief which makes the surrounding rock mass to relax and dilate. Practical experience in the design of rock projects has demonstrated that the estimated rock mass strength of a given slope is affected by the blast damage in slope excavations (Hoek and Brown 1988). In order to improve the accuracy of rock mass strength prediction under disturbance conditions, a disturbance factor D, which varies from zero for undisturbed in situ rock masses to one for highly disturbed rock masses, was introduced to the HB criterion (Eqs. 1.5–1.6) to calculate the HB input parameters. Guidelines for the selection of D will be introduced in Chap. 7.

1.4 Strength and Deformation of Rock Masses Based on the HB Criterion

Rock mass properties, such as uniaxial compressive strength σ_{cm}, deformation modulus E_{m} and shear strength parameters cohesion c and angle of friction ϕ are the most representative parameters to describe the mechanical behavior of rock masses. It is widely applied in numerical modeling of rock engineering projects where the analysis of displacement and stress distribution are required to characterize the rock mass behavior.

Commonly used approaches to estimate rock mass properties include laboratory tests, in situ loading tests and prediction by empirical equations. However, laboratory tests on limited size rock samples containing discontinuities cannot measure reliable values of rock mass properties due to the limitation of the size of the testing equipment (Palmström 1996). In situ tests can provide direct information on the deformability of rock masses, however, as Bieniawski (1973) noted that it is difficult to rely only on one in situ test as different results may be obtained even in a fairly uniform and good rock mass condition. Therefore, in order to obtain reliable results multi-tests are necessary though they are expensive and time-consuming. Due mainly to the above-mentioned difficulties encountered in the laboratory and in situ testing, the estimation of rock mass property values using empirical equations becomes a very attractive and commonly accepted approach among rock engineers.

The uniaxial compressive strength of the rock mass is obtained from the HB criterion by setting σ_3 to zero in Eq. (1.4), given by:

$$\sigma_{cm} = \sigma_{ci} \cdot s^a \tag{1.8}$$

In addition, by putting $\sigma_1 = \sigma_3 = \sigma_{\mathrm{tm}}$ in Eq. (1.4) which represents a condition of biaxial tension, the uniaxial tensile strength σ_{tm} of the rock mass is given as:

$$\sigma_{tm} = -s\sigma_{ci}/m_b \tag{1.9}$$

Hoek (1983) noted that, for brittle rocks, the uniaxial tensile strength is equal to the biaxial tensile strength.

Clearly, similar to its strength, the in situ deformation modulus of a rock mass will depend on the properties of both the intact rock and the discontinuities presented within the rock mass. In the past decades, a great number of empirical equations were proposed for the estimation of the rock mass deformation modulus using various rock mass classification systems, such as the Rock Mass Rating (RMR), the Geological Strength Index (GSI), the Tunneling Quality Index (Q) (Barton 1987, 1996, 2002) and the Rock Mass index (RMi) (Palmström 1996; Palmström and Singh 2001). Empirical models based on GSI are listed in Table 1.1.

In situ data from Bieniawski (1978), Serafim and Pereira (1983) and Stephens and Banks (1989) are gained from high-quality tests and are commonly

Table 1.1 Empirical equations using GSI for predicting deformation modulus of rock masses

Authors	Equations
Hoek and Brown (1997)	$E_m = \sqrt{\frac{\sigma_{ci}}{100}}10^{\left(\frac{GSI-10}{40}\right)}$
Hoek et al. (2002)	$E_m = (1-0.5D)\sqrt{\frac{\sigma_{ci}}{100}}10^{\left(\frac{GSI-10}{40}\right)}, \sigma_{ci} \le 100MPa$ $E_m = (1-0.5D)10^{\left(\frac{GSI-10}{40}\right)}, \sigma_{ci} > 100MPa$
Hoek and Diederichs (2006)	$E_m = E_i\left(0.02 + \frac{1-0.5D}{1+e^{((60+15D-GSI)/11)}}\right)$
Hoek and Diederichs (2006)	$E_m(MPa) = 10^5\left(\frac{1-0.5D}{1+e^{((75+25D-GSI)/11)}}\right)$

acknowledged as reliable data sources. These data were also widely used by many researchers to assess the reliability of their proposed equations. Figure 1.3 gives a comparison among the deformation modulus estimated by Hoek and Diederichs (2006) with D = 0, 0.5 and 1.0. The general agreement among these results suggests that all the predictions of Hoek and Diederichs (2006) can be used with confidence for estimating rock mass deformation modulus.

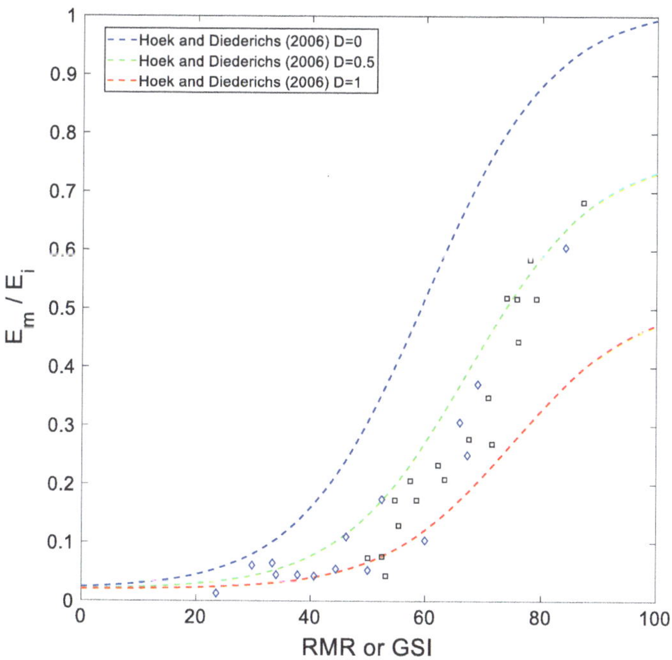

Fig. 1.3 Comparison between field test data and deformation modulus values estimated from Hoek and Diederichs (2006)

1.5 Mohr-Coulomb Parameters Based on the HB Criterion

Because of the analysis methods used in some applications, particularly slope stability analysis, and the needs of the associated software packages, the Mohr-Coulomb shear strength parameters cohesion c and angle of friction ϕ rather than the HB input parameters m_{i}, GSI and D are often required. For example, one of the most popular approaches for estimating the factor of safety (FOS) of a given slope is the limit equilibrium method (LEM) where rock mass strength is usually expressed by the linear MC criterion. If the non-linear HB criterion is used in conjunction with LEM for analyzing the rock slope, methods are required to determine the equivalent MC shear strength parameters, namely cohesion c and angle of friction ϕ, at the specified normal stress σ_{n} from the HB criterion. The determination of reliable shear strength values is a critical step in slope design as small changes in shear strength parameters can result in significant changes in the value of the FOS. In past decades, methods for the determination of shear strength from the Hoek-Brown criterion were proposed by Hoek (1983, 1990), Hoek and Brown (1997), Kumar (1998), Hoek et al. (2002), Carranza-Torres (2004), Priest (2005), Fu and Liao (2010), Shen et al. (2012a, b). A comprehensive review of the literature of estimating shear strength of the Hoek-Brown criterion can be found a paper by Carranza-Torres (2004). However, as Brown (2008) has noted, using the HB criterion to derive exact analytical solutions for estimating the shear strength of a rock mass has proven to be a challenging task due to the complexities associated with mathematical derivation.

The HB criterion Eq. (1.4) can also be expressed in terms of normal stress σ_{n} and shear stress τ on the failure plane calculated by Eqs. (1.10) and (1.11) which were proposed by Balmer (1952).

$$\sigma_n = \sigma_3 + \frac{(\sigma_1 - \sigma_3)}{\partial\sigma_1/\partial\sigma_3 + 1} \tag{1.10}$$

$$\tau = (\sigma_n - \sigma_3)\sqrt{\partial\sigma_1/\partial\sigma_3} \tag{1.11}$$

Taking the derivative of σ_1 with the respect of σ_3 of Eq. (1.4) and substituting the results into Eqs. (1.10) and (1.11) respectively, the HB criterion can be expressed by the following equations:

$$\sigma_n = \sigma_3 + \frac{\sigma_{ci}\left(\frac{m_b\sigma_3}{\sigma_{ci}} + s\right)^a}{2 + am_b\left(\frac{m_b\sigma_3}{\sigma_{ci}} + s\right)^{a-1}} \tag{1.12}$$

$$\tau = (\sigma_n - \sigma_3)\sqrt{1 + am_b\left(\frac{m_b}{\sigma_{ci}}\sigma_3 + s\right)^{a-1}} \tag{1.13}$$

1.5.1 Instantaneous Shear Strength Parameters

The instantaneous MC shear strength parameters can be calculated by locating the tangent of the HB envelope with the specified normal stress σ_n, as illustrated in Fig. 1.4. The slope of the tangent to the HB failure envelope gives an angle of friction ϕ and the intercept of envelope on the shear stress axis gives cohesion c.

Numerical solutions for estimating the equivalent MC shear strength parameters from the HB criterion can be expressed as follows:

$$\phi = \arcsin\left(1 - \frac{2}{2 + am_b\left(\frac{\sigma_3}{\sigma_{ci}}m_b + s\right)^{a-1}}\right) \tag{1.14}$$

$$c = \frac{\sigma_{ci}\cos\phi}{2\left(1 + \frac{\sin\phi}{a}\right)^a}\left(m_b\frac{\sigma_n}{\sigma_{ci}} + s\right)^a - \sigma_n\tan\phi \tag{1.15}$$

$$\frac{\sigma_3}{\sigma_{ci}} = \frac{\sigma_n}{\sigma_{ci}} - \frac{\left(\frac{m_b\sigma_3}{\sigma_{ci}} + s\right)^a}{2 + am_b\left(\frac{m_b\sigma_3}{\sigma_{ci}} + s\right)^{a-1}} \tag{1.16}$$

In the Eqs. (1.14) to (1.16), the values of input parameters m_b, s, a and σ_{ci} are known and normal stress σ_n can be estimated by adopting an appropriate stress analysis approach. In general case of $a \neq 0.5$, there is no analytical solution of Eq. (1.16) for the calculation of the angle of friction ϕ. In order to identify an

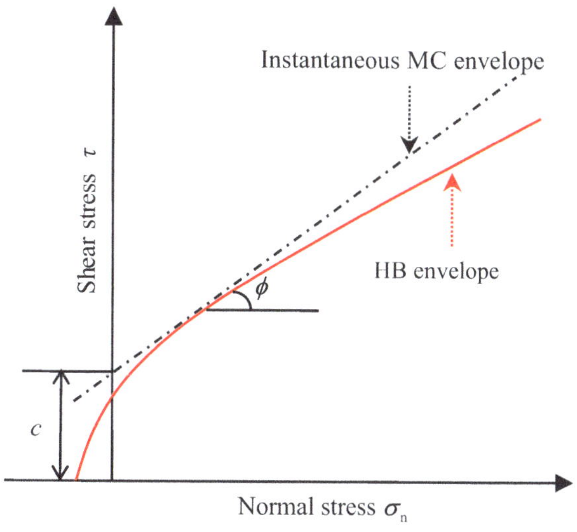

Fig. 1.4 Equivalent MC envelope for the HB criterion

acceptable ϕ value, Eq. (1.16) must be solved numerically. Having obtained ϕ, cohesion c can be calculated by Eq. (1.15).

In special case $a = 0.5$, the analytical solution derived by Bray and reported by Hoek (1983) can yield the accurate MC shear strength parameters for the Hoek-Brown materials. The equations are expressed as follows:

$$h = 1 + \frac{16\left(m_b\sigma_n + s\sigma_{ci}\right)}{3m_b^2\sigma_{ci}} \tag{1.17}$$

$$\theta = \frac{1}{3}\left(90 + \arctan\frac{1}{\sqrt{h^3 - 1}}\right) \tag{1.18}$$

$$\phi = \arctan\frac{1}{\sqrt{4h\cos^2\theta - 1}} \tag{1.19}$$

$$\tau = (\cot\phi - \cos\phi)\frac{m_b\sigma_{ci}}{8} \tag{1.20}$$

$$c = \tau - \sigma_n\tan\phi \tag{1.21}$$

where h and θ are intermediate parameters.

This method provides great flexibility for the use of the original HB criterion ($a = 0.5$) in rock slope stability analysis. However, in the HB criterion, if Eq. (1.7) is used to calculate a, the value of a can vary from 0.51 to 0.5 when GSI is 40 and 100 respectively. Since Bray's method is based on $a = 0.5$ when GSI = 100, the equation gives very good results for rock masses where GSI > 40, as shown in Fig. 1.5. On the other hand, when 0 < GSI < 40, the value of a can vary from 0.666 to 0.51.

Shen et al. (2012a) presented analytical solutions for estimating the Mohr-Coulomb (MC) shear strength of rock masses from the non-linear Generalized Hoek-Brown (HB) criterion using genetic programming (GP). They used Eq. (1.16) to build a GP model as the basis for calculating the intermediate parameter σ_3/σ_{ci} expressed by input parameters m_b, s, a and σ_n/σ_{ci}. After obtaining analytical solution Eq. (1.22) for σ_3/σ_{ci}, closed form solution Eq. (1.25) has been derived for estimating shear stress τ.

$$\frac{\sigma_3}{\sigma_{ci}} = \frac{a\frac{\sigma_n}{\sigma_{ci}}}{\sqrt{a\left(1 + \sqrt{m_b}\right) - \frac{\sigma_n}{\sigma_{ci}}}} \tag{1.22}$$

Substituting Eq. (1.26) into Eq. (1.14), the angle of friction ϕ can be calculated as follows:

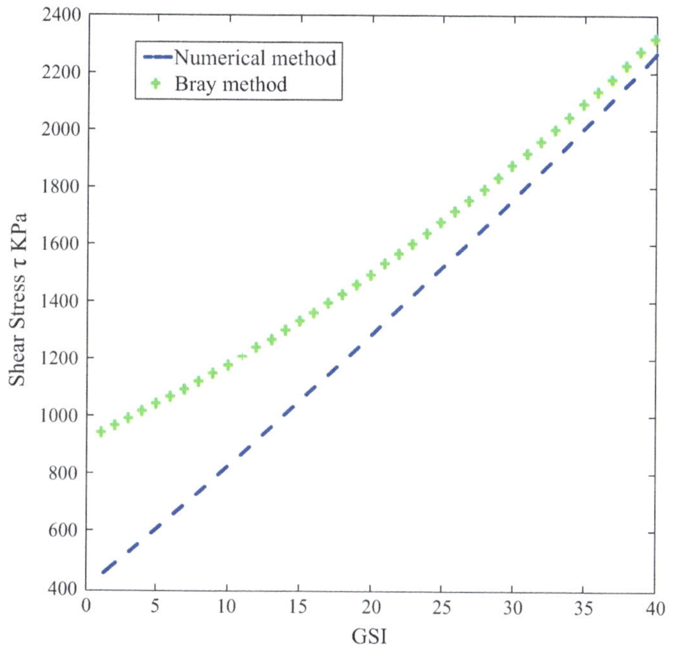

Fig. 1.5 Shear stress versus GSI

$$\phi = \arcsin\left(1 - \frac{2}{P}\right) \tag{1.23}$$

$$P = 2 + am_b\left(m_b\frac{\sigma_3}{\sigma_{ci}} + s\right)^{a-1} \tag{1.24}$$

where P is the intermediate parameter. Finally, the shear stress τ can be expressed as follows:

$$\tau = \sigma_{ci}\frac{\sqrt{P-1}}{P}\frac{\left(m_b\frac{\sigma_n}{\sigma_{ci}} + s\right)^a}{\left(\frac{Pa+P-2}{aP}\right)^a} \tag{1.25}$$

$$c = \sigma_{ci}\frac{\sqrt{P-1}}{P}\frac{\left(m_b\frac{\sigma_n}{\sigma_{ci}} + s\right)^a}{\left(\frac{Pa+P-2}{aP}\right)^a} - \sigma_n\tan\left(\arcsin\left(1 - \frac{2}{P}\right)\right) \tag{1.26}$$

Equation (1.25) is an alternative form of the HB criterion expressed in terms of normal and shear stresses. So that it can be directly used for estimating the instantaneous shear stress of each slice under a specified normal stress in the limit

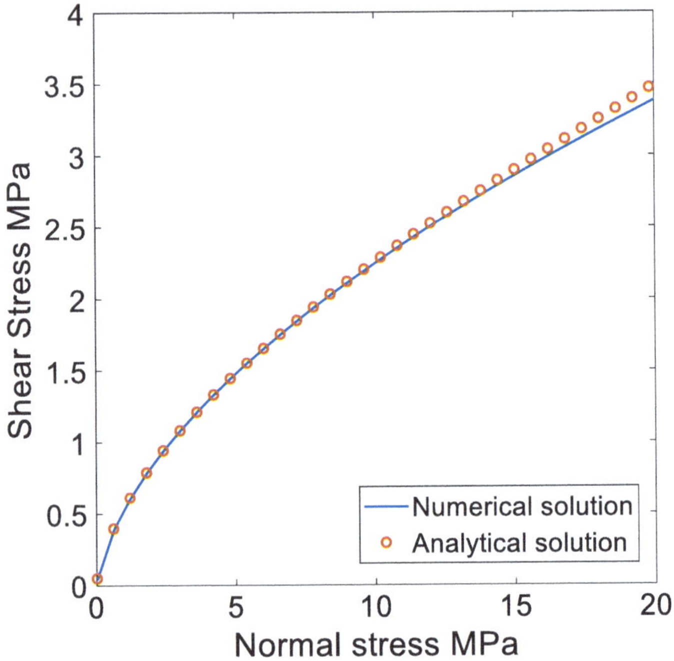

Fig. 1.6 Hoek-Brown shear strength envelope in shear stress/normal stress space (GSI = 20, $m_i = 35$, $D = 1$, $\sigma_{ci} = 30$ MPa)

equilibrium method for the rock slope stability analysis. Figure 1.6 shows the HB envelopes calculated from the analytical and numerical solutions.

The performance of the proposed approximate analytical solution has been tested against numerical solution using 2451 random sets of data. The results show that there is a close agreement between the proposed approximate analytical and numerical solutions. Shear stress τ calculated from the proposed approximate analytical solution exhibits only 0.97% average absolute discrepancy from numerical solutions as shown Fig. 1.7, and the discrepancy of 84.21% sets of data range is less than 2% as shown in Fig. 1.7. In a practical sense, this small difference is acceptable.

1.5.2 Equivalent Shear Strength Parameters

Hoek et al. (2002) proposed an alternative method to estimate equivalent shear strength parameters of rock masses that is using a linear MC envelope to represent the non-linear HB envelope. As illustrated in Fig. 1.8, it is possible to estimate these parameters by fitting a straight line to the HB envelope over a limited range of σ_3, the results are shown in Eqs. (1.27)–(1.32).

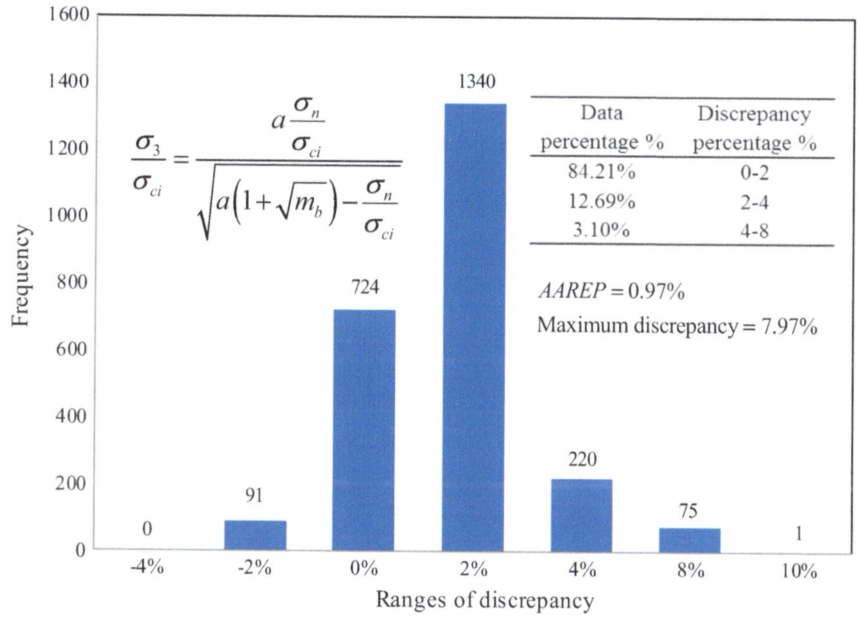

Fig. 1.7 Discrepancy analysis of the proposed analytical solution

Fig. 1.8 Equivalent MC envelope for the HB criterion

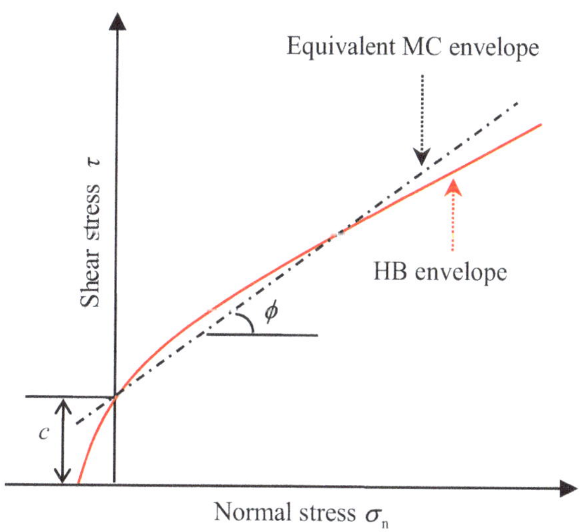

$$\phi = \sin^{-1}\left[\frac{6am_b(s+m_b\sigma_{3n})^{a-1}}{2(1+a)(2+a)+6am_b(s+m_b\sigma_{3n})^{a-1}}\right] \tag{1.27}$$

$$c = \frac{\sigma_{ci}[(1+2a)s+(1-a)m_b\sigma_{3n}](s+m_b\sigma_{3n})^{a-1}}{(1+a)(2+a)\sqrt{1+\left(6am_b(s+m_b\sigma_{3n})^{a-1}\right)/((1+a)(2+a))}} \tag{1.28}$$

$$\sigma_{3n} = \frac{\sigma_{3\max}}{\sigma_{ci}} \tag{1.29}$$

$$\sigma_{3\max} = 0.72\sigma_{cm}\left(\frac{\sigma_{cm}}{\gamma H}\right)^{-0.91} \tag{1.30}$$

(for slope, H is the slope height)

$$\sigma_{3\max} = 0.47\sigma_{cm}\left(\frac{\sigma_{cm}}{\gamma H}\right)^{-0.94} \tag{1.31}$$

(for tunnel, H is the depth of the tunnel below surface)

$$\sigma_{cm} = \frac{\sigma_{ci}(m_b+4s-a(m_b-8s))(m_b/4+s)^{a-1}}{2(1+a)(2+a)} \tag{1.32}$$

By considering m_b, s and a can be calculated from the GSI, D and m_i, therefore, the MC parameters can be calculated from six input parameters (GSI, m_i, D, σ_{ci}, γ, H).

These methods for estimating instantaneous or equivalent shear strength parameters mentioned above are implemented into a software called "RocLab" that can be downloaded from www.rocscience.com.

1.6 The Application of the HB Criterion

After obtaining the HB input data acquired from laboratory tests, geological and excavation condition observations, the rock failure envelope and rock mass properties are calculated based on the HB criterion corresponding to empirical equations. Then, numerical or analytical analyses are carried out to perform a project design to analysis the rock mass displacement, and compared with in-site displacement monitoring results.

A final step is the back analysis of the monitoring results and the feed-back of the results of numerical or analytical analyses. Back analyses are significant in the project since it is the only means whereby the design method and the HB input parameters adopted in the calculations or simulation can be effectively validated. Back analyses should be an ongoing process throughout and even after the construction process of a project so that corrections and adjustments can be made at all stages of the project. Such analysis provides confidence in the design and improves the accuracy of estimating input HB parameters (Hoek and Brown 2018).

References

Balmer G (1952) A general analytical solution for Mohr's envelope. Am Soc Test Mater 52: 1269–1271

Barton N (1987) Rock mass classification, tunnel reinforcement selection using the Q-system. In: Proceedings of the ASTM Symposium on Rock Classification Systems for Engineering Purposes. Cincinnati, Ohio

Barton N (1996) Estimating rock mass deformation modulus for excavation disturbed zone studies. In: International Conference on Deep Geological Disposal of Radioactive Waste, Winnepeg, EDZ workshop. Canadian Nuclear Society, pp 133–144

Barton N (2002) Some new Q value correlations to assist in site characterisation and tunnel design. Int J Rock Mech Min Sci 39:185–216

Bieniawski Z (1973) Engineering classification of rock masses. Trans S African Inst Civ Engrs 15 (12):335–344

Bieniawski Z (1978) Determining rock mass deformability—experience from case histories. Int J Rock Mech Min Sci Geomech Abstr 15:237–247

Brown E (2008) Estimating the mechanical properties of rock masses. In: Proceedings of the 1st southern hemisphere international rock mechanics symposium: SHIRMS 2008, Perth, Western Australia, vol 1, pp 3–21

Carranza-Torres C (2004) Elasto-plastic solution of tunnel problems using the generalized form of the Hoek-Brown failure criterion. Int J Rock Mech Min Sci 2004; 41(3):480–1. In: Hudson JA, Feng X-T, editors. Proceedings of the ISRM SINOROCK 2004 symposium

Fairhurst C (1964) On the validity of the "Brazilian" test for brittle materials. Int J Rock Mech Min Sci 1(4):535–546

Fu W, Liao Y (2010) Non-linear shear strength reduction technique in slope stability calculation. Comput Geotech 37:288–298

Griffith AA (1921)The phenomena of rupture and flow in solids. Philos Trans R Soc Lond (Ser A) 221(2), 163–198

Griffith AA (1924) Theory of rupture. In: Proceedings of the 1st International congress on applied mechanics. Delft, The Netherlands, pp 55–63

Hoek E (1983) Rankine lecture: strength of jointed rock masses. Géotechnique 33:187–223

Hoek E (1990) Estimating Mohr-Coulomb friction and cohesion values from the Hoek-Brown failure criterion. Int J Rock Mech Min Sci Geomech Abstr 27(3):227–229

Hoek E, Brown E (1980a) Underground excavations in rock. Institution of Mining and Metallurgy, London

Hoek E, Brown E (1980b) Empirical strength criterion for rock masses. J Geotech Eng Div 106 (GT9):1013–1035

Hoek E, Brown E (1988) The Hoek-Brown failure criterion: a 1988 update. In: Proceedings of the 15th Canadian Rock Mech. Symp (Edited by Curran J. C.), pp 31–38

Hoek E, Brown E (1997) Practical estimates of rock mass strength. Int J Rock Mech Min Sci 34 (8):1165–1186

Hoek E, Brown E (2018) The Hoeke-Brown failure criterion and GSI-2018 edition. J Rock Mech Geotech Eng. https://doi.org/10.1016/j.jrmge.2018.08.001

Hoek E, Carranza-Torres C, Corkum B (2002) Hoek-Brown criterion e 2002 edition. In: Hammah R, Bawden W, Curran J, Telesnicki M (eds) Mining and tunneling innovation and opportunity, proceedings of the 5th North American rock mechanics symposium and 17th tunnelling association of Canada conference. Toronto, Canada. University of Toronto, Toronto, pp 267–73

Hoek E, Diederichs M (2006) Empirical estimation of rock mass modulus. Int J Rock Mech Min Sci 43:203–215

Hoek E, Martin C (2014) Fracture initiation and propagation in intact rock e a review. J Rock Mech Geotech Eng 6(4):278–300

Kumar P (1998) Shear failure envelope of Hoek-Brown criterion for rockmass. Tunn Undergr Space Technol 13(4):453–458

Mogi K (1966) Pressure dependence of rock strength and transition from brittle fracture to ductile flow. Bull Earthq Res Inst 44:215–232

Palmström A (1996) Characterizing rock masses by RMi for use in practical rock engineering (Part 1: the development of the rock mass index (RMi)). Tunn Undergr Space Technol 11(2):175–188

Palmström A, Singh R (2001) The deformation modulus of rock masses-comparisons between in situ tests and indirect estimates. Tunn Undergr Space Technol 16(3):115–131

Priest S (2005) Determination of shear strength and three-dimensional yield strength for the Hoek-Brown criterion. Rock Mech Rock Eng 38(4):299–327

Read S, Richards L (2014) Correlation of direct and indirect tensile tests for use in the Hoek-Brown constant mi. Rock Engineering and Rock Mechanics: Structures in and on Rock Masses. Taylor and Francis, London

Schwartz A (1964) Failure of rock in the triaxial shear test. In: Proceedings of the 6th rock mechanics symposium. University of Missouri, Rolla, USA, pp 109–151

Serafim J, Pereira J (1983) Considerations on the geomechanical classification of Bieniawski. In: Proceedings of the symposium on engineering geology and underground openings. Lisboa, Portugal, pp 1133–1144

Stephens R, Banks D (1989) Moduli for deformation studies of the foundation and abutments of the Portugues Dam—Puerto Rico. In: Rock mechanics as a guide for efficient utilization of natural resources: Proceedings of the 30th US symposium, Morgantown. Balkema, Rotterdam, pp 31–38

Shen J, Karakus M, Xu C (2012a) Direct expressions for linearization of shear strength envelopes given by the Generalized Hoek-Brown criterion using genetic programming. Comput Geotech 44:139–146

Shen J, Priest SD, Karakus M (2012b) Determination of Mohr-Coulomb shear strength parameters form generalized Hoek-Brown criterion for slope stability analysis. Rock Mech Rock Eng 45:123–129

Ulusay R, Hudson J (2007) The complete ISRM suggested methods for rock characterization, testing and monitoring: 1974-2006. ISRM Turkish National Group, Ankara

Zuo J, Li H, Xie H, Ju Y, Peng S (2008) A nonlinear strength criterion for rocklike materials based on fracture mechanics. Int J Rock Mech Min Sci 45(4):594–599

Zuo J, Liu H, Li H (2015) A theoretical derivation of the Hoek-Brown failure criterion for rock materials. J Rock Mech Geotech Eng 7(4):361–366

Chapter 2
A 2-D Theoretical Derivation of the Hoek-Brown Criterion

Abstract A nonlinear strength criterion for rock-like materials is developed in this chapter. Taking α as an angle of micro-failure orientation in rock-like materials, a formulation between α and load is derived from a mixed-mode fracture criterion based on linear elastic fracture mechanics. According to micro-failure experimental phenomena of rock-like materials, a failure characteristic parameter under triaxial compression condition is chosen, which is relevant to confining pressure and is an invariant. A theoretical nonlinear strength criterion is also derived, which is exactly in the same mathematical form as the original Hoek–Brown empirical strength criterion. In addition, it is also found that the coefficient m in the Hoek–Brown criterion has physical meaning which is related to the ratio between the uniaxial compressive strength and the uniaxial tensile strength.

2.1 Introduction

Since the problem of strength was firstly discussed by Leonardo da Vinci (Williams 1957), a considerable amount of theoretical and experimental research on strength under complex stress states have been conducted for various materials. To meet the increasing demand for rock engineering, civil engineering, and underground excavation, many strength criteria for rock-like materials have also been developed, such as the Griffith criterion (1921), Mohr–Coulomb theory (Andreev 1995), and Hoek–Brown failure criterion (1980a, b, 1983, 1997), Hoek et al. (1992), Hoek (1998). The Hoek–Brown criterion is indeed the most widely applied one in rock engineering.

There have been classically two different approaches, i.e. macroscopic and micro/mesoscopic methods, to develop a theory of brittle failure for rock-like materials that are used to predict the macroscopic fracture stress (Wiebols and Cook 1968; Brady 1969; Gregg 1985; Wang et al. 2005; Paterson and Wong 2005; Gong et al. 2019; Si et al. 2019). The macroscopic methods put more emphasis on the growth of the crack and the macro-failure characteristics, but they say very little about the physical mechanisms of failure, such as the Mohr–Coulomb criterion. Considering the loading rate effect, a dynamic empirical Mohr-Coulomb criterion expression of marble under low loading

J. Zuo and J. Shen, *The Hoek-Brown Failure criterion—From theory to application*,
https://doi.org/10.1007/978-981-15-1769-3_2

rates ($10^{-5} \sim 10^{-3}$ MPa•s^{-1}) by Gong et al. (2018). Further more, Gong et al. (2019) and Si et al. (2019) conducted the dynamic uniaxial and triaxial compression tests of sandstone at different confining pressures and high strain rate (40 s^{-1} to 160 s^{-1}) using a conventional and modified triaxial SHPB system, and then proposed the unified dynamic Mohr-Coulomb and Hoek-Brown strength criteria at high strain rates respectively. These criteria are given in terms of stresses σ_1 and σ_3 without considering the intermediate principal stress σ_2. However, evidence has shown that σ_2 does influence the failure (Al-Ajmi and Zimmerman 2005). The micro/mesoscopic method, which puts more emphasis on micro-mechanism of rock failure, attempts to set up a physical model of the brittle fracture process that is amenable to theoretical treatment. Although a certain degree of empiricism may still be involved, these models are intended to represent essential aspects of the actual physical mechanism of fracture and provide a firmer theoretical basis for establishing criteria of failure which are applicable to general states of stress. Though these rock strength criteria have paid attention to the initiation and propagation of critical microcracks (Wiebols and Cook 1968; Brady 1969), little consideration has been given to the effect that the confining pressure may have in restraining the microcrack propagation. In addition, there is unfortunately a great discrepancy between theoretically predicted values and experimental values. Therefore, it is necessary to seek a new characteristic parameter and strength criterion to depict rock-like material failure. In this paper, the brittle micro/meso failure processes of rock-like materials under compressive loading are fully investigated, and a theoretical nonlinear strength criterion for rock-like materials is presented based on linear elastic fracture mechanics.

2.2 Fundamental Hypothesis and 2-D Model

Griffith's theory of brittle failure is based on the assumption that the low order of strength in brittle materials is due to the presence of microcracks or flaws. There is abundant evidence for the existence of crack-like flaws in many brittle materials (Paterson and Wong 2005; Jaeger et al. 2007). Therefore, Griffith's model is physically plausible, and it has been intensively studied and developed to derive macroscopic criteria of failure (Paterson and Wong 2005; Jaeger et al. 2007). In this paper, Griffith's crack is adopted, and the following assumptions are made: (1) numerous Griffith microcracks are distributed randomly in the rock sample; (2) these microcracks do not interact with each other; (3) the matrix material in which microcracks are embedded is isotropic; and (4) these microcracks are closed.

The heterogeneous nature of rock gives rise to various types of flaws or local stress concentrations that might act as sources for growing microcracks (Kemeny and Cook 1991). On the basis that the most likely sources are the initially present microcracks, the micro-mechanical model that has been analyzed most thoroughly is that of the "sliding crack" (Brace and Bombolakis 1963; Brace et al. 1966; Fairhurst and Cook 1966; Kachanov 1982a, b). Because these microcracks are assumed not to interact mutually, a model considers a pre-existing sliding crack in an arbitrary plate of unit thickness with the initial length $2a$ and orienting at angle β to the direction of major principal stress σ_1, is shown in Fig. 2.1.

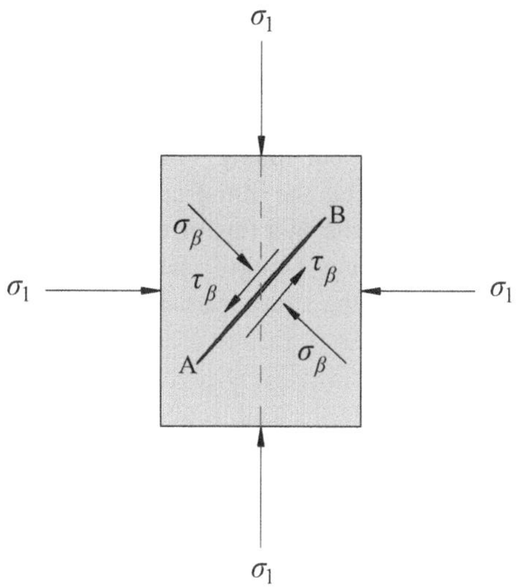

Fig. 2.1 Mechanical model for an initial sliding crack in the rock

2.3 Critical Crack Propagation Analysis

According to fracture mechanics (Anderson 2005), in-plane shear stress causes a deformation of the sliding crack such that the displacement of the crack surfaces is in the plane of the crack and perpendicular to the leading edge of the crack. Then the effective shear stress τe can be developed as follows:

$$\tau_e = \tau_\beta - \tau_f \tag{2.1}$$

where τ_e is the effective shear stress, τ_β is the tangential stress of the crack, τ_f is the frictional shear stress on plane AB (Fig. 2.1). $\tau_f = \mu\sigma_\beta$, μ is the friction coefficient, σ_β is the normal stress acting on plane AB, and β is the orientation angle between the failure surface and the direction of major principal stress σ_1. Applying the stress transformation equations, the normal stress σ_β, and tangential shear stress τ_β are given by

$$\sigma_\beta = \frac{\sigma_1 + \sigma_3}{2} - \frac{\sigma_1 - \sigma_3}{2}\cos 2\beta \tag{2.2}$$

$$\tau_\beta = \frac{1}{2}(\sigma_1 - \sigma_3)\sin 2\beta \tag{2.3}$$

Substituting Eqs. (2.2) and (2.3) into Eq. (2.1), the effective shear stress τ_e can be given by

$$\tau_e = \frac{1}{2}[(\sigma_1 - \sigma_3)(\sin 2\beta + \mu \cos 2\beta) - \mu(\sigma_1 + \sigma_3)] \tag{2.4}$$

For the typical two-dimensional problem (Fig. 2.1), the effective shear stress τ_e initiates frictional sliding. And if τ_e is sufficiently high to overcome the frictional resistance along the initial crack, frictional slip that results in tensile stress concentration at the two tips of the sliding crack, which, in turn, may induces nucleation of "wing cracks", as shown in Fig. 2.2. Prior to the onset of crack propagation, the stress intensity factor at the initial crack tip is (Tada 1973):

$$K_{\mathrm{II}} = \tau_e \sqrt{\pi a} \tag{2.5}$$

If the wing crack length is infinitesimal, the stress intensity factor at the tip of the secondary crack based on linear elastic fracture mechanics can be expressed by Kachanov (1982b), Cotterell and Rice (1963):

$$K_{\mathrm{II}} \geq \kappa K_{\mathrm{I}C} \tag{2.6}$$

where κ can be derived from the various approximations suggested in the literature on the kinked crack, such as $\kappa = \sqrt{3}/2$ in maximum-stress criterion (Sih and Macdonald 1974), $\kappa = \sqrt{3(1-v)/(2+2v-v^2)}$ (where v is Poisson's ratio) in the minimum-strain-energy criterion (Sih 1974; Nuismer 1975), and $\kappa = 1$ in the maximum energy release-rate criterion (Diederichs 2003). Fracture toughness K_{IC} can be given by induced tensile strength σ_t and crack length a, namely:

$$K_{\mathrm{I}C} = \sigma_t \sqrt{\pi a} \tag{2.7}$$

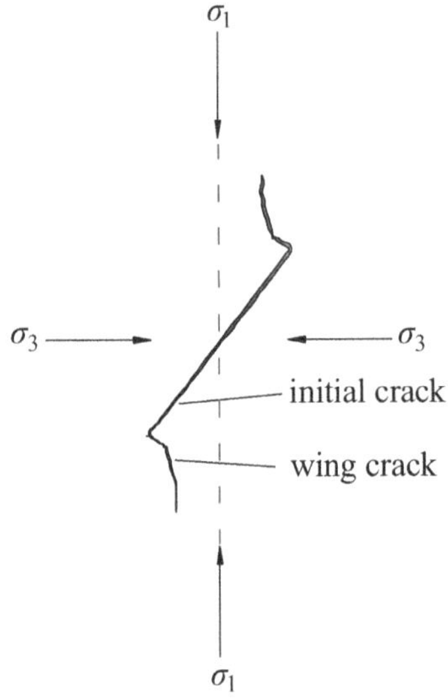

Fig. 2.2 Propagation of wing cracks from the tip of an initial crack

Substituting Eqs. (2.5) and (2.7) into Eq. (2.6), then:

$$\tau_e \geq \kappa\sigma_t \tag{2.8}$$

Let the orientation angle β of the critical propagating crack in rock-like material satisfies the following two equations:

$$\frac{\partial K_{\text{II}}}{\partial \beta} = \sqrt{\pi \text{a}}\frac{\partial \tau_e}{\partial \beta} = 0 \tag{2.9}$$

$$\frac{\partial^2 K_{\text{II}}}{\partial \beta^2} < 0 \tag{2.10}$$

Then we can obtain the critical orientation angle β_O using Eqs. (2.4), (2.5), and (2.9):

$$\beta_0 = \frac{1}{2}\arctan(1/\mu) \tag{2.11}$$

2.4 Micro-failure Orientation Angle α of Rock

Many wing cracks will nucleate, initiate and propagate at the tip of initial cracks when the rock is subjected to compression. If we converge the lower and points of all those wing cracks, which meet with Eq. (2.8), to point **O**, then we can obtain a fan-shaped area of crack distribution zone with a radius of *2a*. The included angle α of the fan section is defined as the "micro-failure orientation angle" in which all of the fitting microcracks have occupied, as shown in Fig. 2.3. And the micro-failure orientation angle should be a monotonic increasing function of micro-failure. Here, two dimensionless parameters are defined: $\sum_1 = \sigma_1/\sigma_t$ and $\sum_3 = \sigma_3/\sigma_t$. Substituting Eq. (2.4) into Eq. (2.8), we have

$$\left(\mu\sum\nolimits_1 + \kappa\right)\tan^2\beta - \left(\sum\nolimits_1 - \sum\nolimits_3\right)\tan\beta + \left(\mu\sum\nolimits_3 + \kappa\right) \leq 0 \tag{2.12}$$

From the inequality, the tangent of β can be determined by

$$\tan\beta_1 \leq \tan\beta \leq \tan\beta_2 \tag{2.13}$$

where

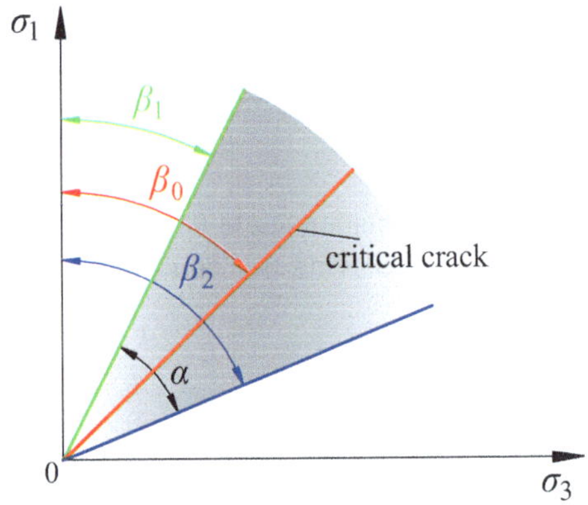

Fig. 2.3 Crack distribution zone

$$\tan\beta_1 = \frac{(\sum_1 - \sum_3) - \sqrt{(\sum_1 - \sum_3)^2 - 4(\mu\sum_1 + \kappa)(\mu\sum_3 + \kappa)}}{2(\mu\sum_1 + \kappa)} \tag{2.14}$$

$$\tan\beta_2 = \frac{(\sum_1 - \sum_3) - \sqrt{(\sum_1 - \sum_3)^2 - 4(\mu\sum_1 + \kappa)(\mu\sum_3 + \kappa)}}{2(\mu\sum_1 + \kappa)} \tag{2.15}$$

Using Eqs. (2.14) and (2.15), we can obtain:

$$\tan(\beta_1 + \beta_2) = \frac{\tan\beta_1 + \tan\beta_2}{1 - \tan\beta_1\tan\beta_2} = \frac{1}{\mu} = \tan 2\beta_0 \tag{2.16}$$

i.e.:

$$\beta_1 + \beta_2 = 2\beta_0 \tag{2.17}$$

Transforming the equation form, we can say that:

$$\beta_0 - \beta_1 = \beta_2 - \beta_0 \tag{2.18}$$

Equation (2.18) indicates that the propagation angle domain of microcracks is symmetrical in the critical crack orientation. Supposing, $\alpha = \beta_2 - \beta_1$, we have

$$\begin{aligned}\tan\alpha = \tan(\beta_2 - \beta_1) &= \frac{\tan\beta_2 - \tan\beta_1}{1 + \tan\beta_1\tan\beta_2}\\ &= \frac{\sqrt{(\sum_1 - \sum_3)^2 - 4(\mu\sum_1 + \kappa)(\mu\sum_3 + \kappa)}}{\mu(\sum_1 + \sum_3) + 2\kappa}\end{aligned} \tag{2.19a}$$

2.5 Rock Failure Characteristic Parameter

Though the brittle failure of rock is very complicated under loading, the failure characteristic parameter should be an invariant when the rock breaks entirely. Brady (1969) has suggested that failure occurs when the volumetric strain attains a critical value due to microcracking. A large number of rock experiments have also revealed that the initiation of an internal microcrack does not indicate rock failure (Paterson and Wong 2005). Therefore, crack initiation cannot be chosen as a brittle failure characteristic. Jaeger et al. (2007) have put forward that local transverse tensile stresses can arise at a microcrack by considering the two-dimensional analysis of stresses around an elliptical hole or crack. Then we can expect local transverse tensile stresses to arise at the ends of any axially oriented microcracks in brittle rock if the confining pressure is not too high, thus favoring the propagation of crack parallel to the specimen axis from the microcrack. The microcrack will tend not to propagate in its own plane but swing into an orientation more nearly parallel to the orientation of major principal stress σ_1, as shown in Fig. 2.2. Many researchers, for instance (Fairhurst and Cook 1966), have therefore used the induced tensile stresses to explain why cracks of tension propagate parallel to the direction of compressive loading. Then, this problem becomes a stability problem. To assure that the wing crack propagates continually, the major principal stress σ_1 should be increased further. In addition, the confining pressure σ_3 can effectively restrain wing micro-cracks from propagating arbitrarily. Therefore, the initiation of cracks cannot indicate rock failure. Rock specimen fractures when internal micro-failure density reaches a critical value, but the value is not a material constant, instead it is related to external load and rock specimen. Moreover, confining pressure does play a part in restraining microcracks propagation (Paterson and Wong 2005), namely, an increase in confining pressure σ_3 leads to a decrease in the micro-failure density. Therefore, rock failure characteristic parameter should be relevant to confining pressure. Also, it relates to the micro-failure orientation angle α. Therefore, the failure characteristic value should satisfy the following three principles: firstly, the higher the confining pressure σ_3 is when rock fractures, the lower the micro-failure orientation angle α will be; secondly, the result from the selected characteristic parameter should be in a simple mathematic expression; finally, the result should agree well with the experimental data. According to the first principle: the higher the confining pressure σ_3 is, the lower the micro-failure orientation angle α is, and then the following relationship can be concluded:

$$\begin{cases} 0 < \overset{1}{\sum}{}_3 < \overset{2}{\sum}{}_3 < \overset{3}{\sum}{}_3 < \cdots < \overset{n}{\sum}{}_3 \\ \phi_0 > \phi_1 > \phi_2 > \phi_3 > \cdots > \phi_n \\ \cos\phi_0 < \cos\phi_1 < \cos\phi_2 < \cos\phi_3 < \cdots < \cos\phi_n \end{cases} \tag{2.20}$$

where a dimensionless parameter of $\sum^{i} 3$ (i = 0, 1, 2, …, n) at the moment of rock failure corresponds to a micro-failure orientation angle $\overset{i}{\alpha}$ (i = 0, 1, 2, …, n) and the dimensionless parameter $\sum^{i} 3 = 0$.

Obviously, the expression of the micro-failure orientation angle α is complicated and cannot be selected as the failure characteristic parameter. According to the second principle and Eq. (3.19), the cosine of the micro-failure orientation angle α can be written as:

$$\cos\alpha = \frac{\mu}{\sqrt{1+\mu^2}}\left(1 + 2\frac{\sum_3 + \kappa/\mu}{\sum_1 - \sum_3}\right) \tag{2.21}$$

For an invariable confining pressure, namely invariable dimensionless parameter $\sum 3$, we can find the relationship between cos α and the dimensionless parameter $\sum 1$. Then we can obtain the relationship curve of cos α and the dimensionless parameter $\sum 3$ because μ and κ both are constant, as shown in Fig. 2.4. The bold curve denotes the failure curve. The relationship indicates that the dimensionless parameter $\sum 1$ and the cosine of the micro-failure orientation angle α both increase with the dimensionless parameter $\sum 3$, namely, both σ_1 and $\cos\alpha$ increase with the increment of σ_3.

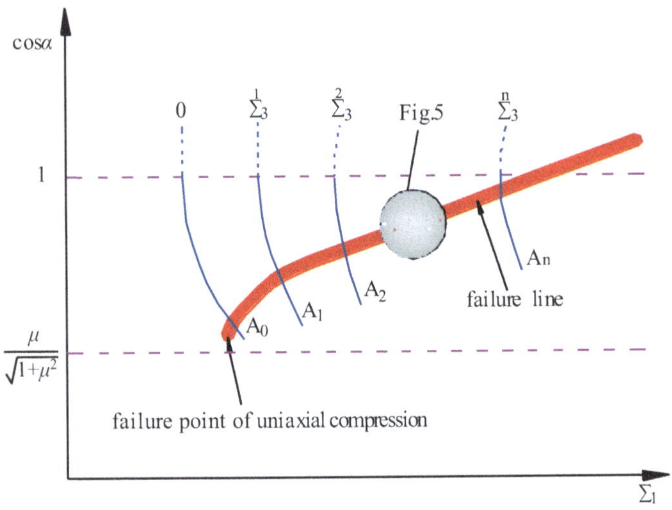

Fig. 2.4 cos α with $\sum_1$ for different $\sum_3$

2.6 A Theoretical Derivation of the Hoek-Brown Failure Criterion

From Eq. (2.21), we can obtain

$$\frac{1}{\left(\sum_1 - \sum_3\right)^2} = \left[\frac{(\sqrt{1+\mu^2}\cos\alpha)/\mu - 1}{2\sum_3 + 2\kappa/\mu}\right]^2 \tag{2.22}$$

Differentiating Eq. (2.21) with respect to $\sum_1$, we obtain

$$\left|\frac{\partial\cos\alpha}{\partial\sum_1}\right| = \frac{\mu}{\sqrt{1+\mu^2}}\frac{2\sum_3 + 2\kappa/\mu}{\left(\sum_1 - \sum_3\right)^2} \tag{2.23}$$

where $|\partial\cos\alpha/\partial\sum 1|$ is defined as the rate of change of cos α to the axial stress dimensionless parameter $\sum_1$. Substituting Eq. (2.22) into Eq. (2.23), then it gives:

$$\left|\frac{\partial\cos\alpha}{\partial\sum_1}\right| = \frac{\mu}{\sqrt{1+\mu^2}}\frac{\left[(\sqrt{1+\mu^2}\cos\alpha)/\mu - 1\right]^2}{2\sum_3 + 2\kappa/\mu} \tag{2.24}$$

An arbitrary failure point A_i on the failure curve is enlarged as shown in Fig. 2.5. A_i corresponds to the dimensionless parameter $\overset{i}{\sum} 3$ and another failure point A_{i+1} corresponds to $\overset{i+1}{\sum} 3$. The dimensionless parameter $\overset{i}{\sum} 3$ is less than $\overset{i+1}{\sum} 3$. From the geometrical relationship shown in Fig. 2.5, the movement process from A_i to A_{i+1} along the failure curve can be divided into two segments. Segment one is from A_i to A_{i+1} where cos α is constant and $\sum_3$ increases, and segment two is from A'_{i+1} to A_{i+1} where cos α increases and $\sum_3$ stays constant. From Eq. (2.24) and Fig. 2.5, the increase of dimensionless parameter $\sum_3$ results in the decrease of $|\partial\cos\alpha/\partial\sum 1|$ in segment one. But in segment two, the increase of cos α results in the increase of $|\partial\cos\alpha/\partial\sum 1|$. In the two segments, if the increment of $|\partial\cos\alpha/\partial\sum 1|$ in segment two equals to the decrement of $|\partial\cos\alpha/\partial\sum 1|$ in segment one, then the rate of change $|\partial\cos\alpha/\partial\sum 1|$ is always a constant when the rock fractures, no matter what the value of $\sum_3$ is. Therefore, the rate of change $|\partial\cos\alpha/\partial\sum 1|$ can be regarded as the rock failure characteristic parameter, and it can satisfy the first principle. Also, $|\partial\cos\alpha/\partial\sum 1|$ could possibly be a strength criterion when rock is subjected to loading. Through the uniaxial compression conditions $\sum_1 = \sigma_c/\sigma_t = \sum_c$ and $\sum_3 = 0$, we can obtain the rate of change constant $|\partial\cos\alpha/\partial\sum 1|$, namely,

$$\left|\frac{\partial\cos\alpha}{\partial\sum_1}\right| = \frac{\mu}{\sqrt{1+\mu^2}}\frac{2\kappa}{\mu\sum_c^2} \tag{2.25}$$

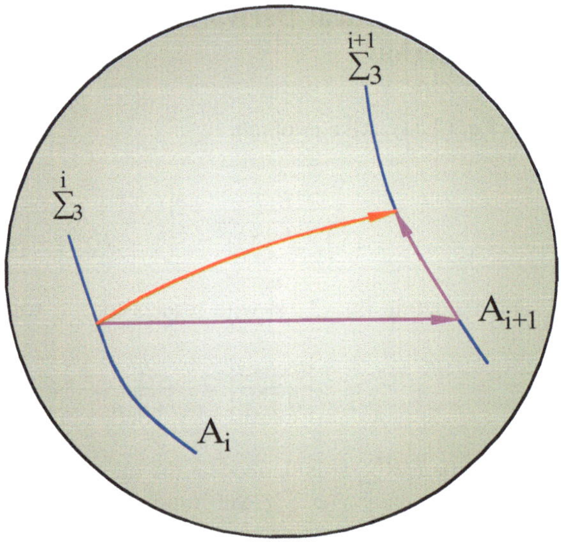

Fig. 2.5 cos α with $\sum_1$ for different $\sum_3$ (locally enlarged view)

From Eqs. (3.23) and (3.25), the following equation is derived:

$$\sum{}_1 = \sum{}_3 + \sqrt{\frac{\mu}{\kappa}\sum{}_c^2 \sum{}_3 + \sum{}_c^2} \tag{2.26}$$

Equation (2.26) indicates that the dimensionless parameter $\sum_1$ is the monotonic increasing function of the confining stress dimensionless parameter $\sum_3$. This relationship is also consistent with experimental data. The authors found that the differential coefficient of α or other trigonometric functions of α with respect to dimensionless parameter $\sum_1$, except $|\partial \sec\alpha/\partial \sum 1|$, cannot be chosen as the failure characteristic parameter because they do not satisfy the first principle. Thus, not choosing $|\partial \sec\alpha/\partial \sum 1|$ as the failure characteristic parameter is in that the dimensionless parameter $\sum_1$ is not the monotonic increasing function of the dimensionless parameter $\sum_3$. This does not agree with experimental data.

Substituting the dimensionless parameters $\sum_1 = \sigma_1/\sigma_t$ and $\sum_3 = \sigma_3/\sigma_t$ into Eq. (2.26), then

$$\sigma_1 = \sigma_3 + \sqrt{\frac{\mu}{\kappa}\frac{\sigma_c}{\sigma_t}\sigma_c\sigma_3 + \sigma_c^2} \tag{2.27}$$

This nonlinear theory strength criterion for rock-like materials has been derived theoretically from fracture mechanics, which is based on micro-failure phenomena. We see that Eq. (2.27) is remarkably similar to the original Hoek–Brown empirical criterion (1980a, b):

$$\sigma_1 = \sigma_3 + \sqrt{\mathrm{m}\sigma_c\sigma_3 + s\sigma_c^2} \tag{2.28}$$

With $s = 1$ for intact rock, then

$$\sigma_1 = \sigma_3 + \sqrt{\mathrm{m}\sigma_c\sigma_3 + \sigma_c^2} \tag{2.29}$$

Taking m to replace $(\mu/\kappa)(\sigma_c/\sigma_t)$, Eq. (2.27) then has the same mathematical form with Eq. (3.29) for intact rock. And this indicates that m is related to the friction coefficient μ, the coefficient κ of mixed-mode fracture criterion, and the uniaxial compressive strength σ_c and uniaxial tensile strength σ_t. Therefore, m can characterize rock strength.

References

Al-Ajmi AM, Zimmerman RW (2005) Relation between the Mogi and the Coulomb failure criteria. Int J Rock Mech Min Sci 42(3):431–439

Anderson TL (2005) Fracture mechanics: fundamentals and applications. Taylor & Francis, Boca Raton, FL

Andreev GE (1995) Brittle failure of rock materials: test results and constitutive models. Balkema, Rotterdam

Brace WF, Bombolakis EG (1963) A note on brittle crack growth in compression. J Geophys Res 68:3709–3713

Brace WF, Paulding BW, Scholz C (1966) Dilatancy in the fracture of crystalline rocks. J Geophys Res 71:3939–3953

Brady BT (1969) A statistical theory of brittle fracture for rock materials. Part I: brittle failure under homogeneous axisymmetric states of stress. Int J Rock Mech Min Sci 6:21–42

Cotterell B, Rice JR (1963) Slightly curved or kinked cracks. Int J Fract 1980, 16(2):155–169. Erdogan F, Sih GC On the crack extension in plates under plane loading and transverse shear. J Basic Eng 85:519–527

Diederichs MS (2003) Rock fracture and collapse under low confinement conditions. Rock Mech Rock Eng 36(5):339–381

Fairhurst C, Cook NGW (1966) The phenomenon of rock splitting parallel to the direction of maximum compression in the neighbourhood of a surface. In: Proceedings of the first congress of the international society of rock mechanics, Lisbon, pp 687–692

Gong FQ, Luo S, Lin G, Si XF (2018) Dynamic empirical Mohr–Coulomb criterion of marble under low loading rate. In: Proceeding of International conference on geo-mechanics, geo-energy and geo-resources, IC3G, Chengdu, China, 22–24 Sept 2018, pp. 87–95

Gong FQ, Si XF, Li XB, Wang SY (2019) Experimental investigation of strain rockburst in circular caverns under deep three-dimensional high-stress conditions. Rock Mech Rock Eng 52 (5):1459–1474

Gregg WJ (1985) Microscopic deformation mechanisms associated with mica film formation in cleaved Psammitic rocks. J Struct Geol 7:45–56

Griffith AA (1921) The phenomena of rupture and flow in solids. Philos Trans Roy Soc London Ser A. 221:163–198

Hoek E (1998) Reliability of Hoek–Brown estimates of rock mass properties and their impact on design. Int J Rock Mech Min Sci 35(1):63–68

Hoek E, Brown ET (1980a) Underground excavations in rock. Institute of Mining and Metallurgy, London

Hoek E, Brown ET (1980b) Empirical strength criterion for rock mass. J Geotech Eng Div ASCE. 106(9):1013–1035

Hoek E, Brown ET (1983) Strength of jointed rock masses. Geotechnique 33:187–223

Hoek E, Brown ET (1997) Practical estimates of rock mass strength. Int J Rock Mech Min Sci 34:1165–1186

Hoek E, Wood D, Shah S (1992) A modified Hoek–Brown criterion for jointed rock masses. In: Hudson JA (ed) Proceedings of the rock characterization, ISRM symposium Eurock'92. British Geotechnical Society, London, pp 209–214

Jaeger JC, Cook NGW, Zimmerman RW (2007) Fundamentals of rock mechanics, 4th edn. Blackwell, Oxford

Kachanov ML (1982a) A microcrack model of rock inelasticity, Part I: frictional sliding on microcracks. Mech Mater 1:19–27

Kachanov ML (1982b) A microcrack model of rock inelasticity, Part II: propagation of microcracks. Mech Mater 1:29–41

Kemeny JM, Cook NGW (1991) Micromechanics of deformation in rocks. In: Shah SP (ed) Toughening mechanisms in quasi-brittle materials. Kluwer Academic Publishers, Dordrecht, pp 155–188

Nuismer RJ (1975) An energy release rate criterion for mixed mode fracture. Int J Fract 11(2):245–250

Paterson MS, Wong TF (2005) Experimental rock deformation—the brittle field. Springer, Berlin

Si XF, Gong FQ, Li XB, Wang SY, Luo S (2019) Dynamic Mohr–Coulomb and Hoek–Brown strength criteria of sandstone at high strain rate. Int J Rock Mech Min Sci 115:48–59

Sih GC (1974) Strain-energy-density factor applied to mixed mode crack problems. Int J Fract 10 (3):305–321

Sih GC, Macdonald B (1974) Fracture mechanics applied to engineering problems-strain-energy density fracture criterion. Eng Frac Mech 6(2):361–386

Tada H (1973) Stress analysis of cracks handbook. Del Research Corporation, Hellertown, PA

Wang XS, Wu BS, Wang QY (2005) Online SEM investigation of microcrack characteristics of concretes at various temperatures. Cement Concr Res 35:1385–1390

Wiebols GA, Cook NGW (1968) An energy criterion for the strength of rock in polyaxial compression. Int J Rock Mech Min Sci 5:529–549

Williams E (1957) Some observations of Leonardo, Galileo, Mariotte and others relative to size effect. Ann Sci 13:23–29

Chapter 3
A 3-D Theoretical Derivation of the Hoek-Brown Criterion

Abstract A Hoek-Brown rock failure criterion is derived by a three-dimensional crack model based on the linear elastic fracture theory. We believe that when macro-failure occurs, the failure characteristic factor needs to exceed the critical value. This factor is a product of the micro-failure orientation angle and the changing rate of the angle with respect to the major principal stress. And the factor mathematically leads to the empirical Hoek-Brown rock failure criterion is further discussed. Thus, the proposed factors can successfully relate the evolution of micro-damage crack characteristics to macro-failure. A quantitative relationship between the brittle-ductile transition point and confining pressure is proposed based on the above theoretical development, which is consistent with experimental observations.

3.1 Introduction

Failure of brittle materials, such as rock and concrete, has been a subject of intensive study for many years in rock physics and geotechnical engineering. Many mechanical and material science workers have been concerning largely with the failure of brittle material in their attempts to design and evaluate the safety and stability of engineering structure. Therefore, the failure criterion for brittle material is unquestionably one of the most important subjects (Hoek and Brown 1980; Hoek and Brown 1997; Nemat-Nasser and Hori 1993; Yu 2004; Paterson and Wong 2005; Jaeger et al. 2007; Zuo et al. 2008). Numerous macro/meso/microscopic experiments have been carried out to investigate the failure mechanism of brittle materials under different stresses and thermal conditions (Nemat-Nasser and Hori 1993; Yu 2004; Paterson and Wong 2005; Jaeger et al. 2007; Wang et al. 2012a, b; Jian-ping et al. 2011; Zuo Jian-ping et al. 2010). In addition, some corresponding numerical simulation methods have also been developed to study materials failure behavior (Li et al. 2011, 2012; Pan et al. 2012a, b). However, most of the literature have focused on the failure of materials containing single crack or a limited number of cracks. How to evaluate the brittle failure behavior of materials containing a large

J. Zuo and J. Shen, *The Hoek-Brown Failure criterion—From theory to application*,
https://doi.org/10.1007/978-981-15-1769-3_3

number of random distributions of micro-cracks still has no good solution. In this paper, new insight has been focused on rock final failure including numerous internal microfracturing. And a nonlinear strength criterion can also be derived based on a sliding crack model at three-dimensional space. A theoretical study of brittle fracture aims to explain both the phenomenological and laboratory observations. In addition, the criterion has been applied to investigate the phenomena of brittle-ductile transition in rocks at last.

3.2 Fundamental 3-D Model for Cracked Brittle Materials

Since the first works of (Griffith 1921), there is abundant evidence for the existence of Griffith's cracks or other crack-like flaws in brittle materials (Griffith 1921; Peng and Johnson 1972; Schovanec 1986; Mura 1987; Vardoulakis and Papamichos 1991; Golshani et al. 2006). In addition, the low strength in some brittle materials is mainly due to the presence of these micro cracks or flaws. In the present work, we assumed that numerous Griffith micro-cracks distribute randomly in rocks. The model of frictional sliding crack has been widely discussed in many kinds of literature (Holcomb 1978; Kachanov 1982a, b; Le and Yang 2006). Although, the crack interactions can be taken into account by modifications of a self-consistent method, in this paper, these micro-cracks interaction would be neglected so that tractions on crack faces can be calculated as one that induced by an external load. The effects of the shape and size of micro-cracks on rock property are not considered. Therefore, a simplified model for a sliding disk-shaped crack with diameter of $2a$ in an infinite plate is shown in Fig. 3.1. We assume that it is a closed crack. The sign convention adopted for stress is that compressive stress is positive and tensile stress is negative. The maximum, intermediate, and minimum principal stresses are denoted by σ_1, σ_2, σ_3 respectively, and the three subscripts represented the coordinate axes x_1, x_2, x_3 respectively. The orientation of disk-shaped crack can be defined by direction cosine l, m, n. If a unique sphere is built up with the equations

$$l^2 + m^2 + n^2 = 1 \tag{3.1}$$

then the orientation of the crack plane is determined by the normal point **M**. Therefore, the normal stress σ_n and shear stress τ_n components on the disk-shaped crack are given in terms of the three principal stresses and as a function of direction cosine l, m, n:

$$\sigma_n = l^2\sigma_1 + m^2\sigma_2 + n^2\sigma_3 \tag{3.2}$$

$$\tau_n^2 = l^2\sigma_1^2 + m^2\sigma_2^2 + n^2\sigma_3^2 - \sigma_n^2 \tag{3.3}$$

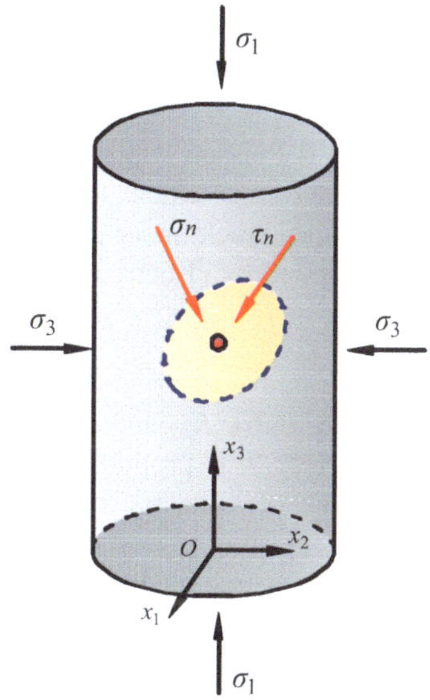

Fig. 3.1 A disk-shaped crack of diameter 2a in an infinite plate under triaxial compression

3.3 Propagation Condition of Disk-Shaped Crack

Although it is acknowledged that non-elastic effects are involved at crack tips in various rocks, and that even the elastic behavior in the most highly stressed regions may be nonlinear, the practical analysis of the stress distribution in the neighborhood of the crack tip is usually done on the basis of the classical linear theory of elasticity. Therefore, according to linear elastic fracture mechanics (Anderson 2005), the mode of crack in Fig. 3.1 is in-plane shear mode (mode II). The mode corresponds to in-plane shear loading and tends to slide one crack face with respect to the other. The stress intensity factor depends both on shear stress τ_n and friction stress τ_μ which are parallel to the crack plane. According to Amonton's law, $\tau_\mu = \mu\sigma_n$ where μ and σ_n are the friction coefficient for the pre-existing sliding crack surface and the normal stress acting on the disk plane, respectively. It is obvious that τ_μ is a key factor which can suppress crack propagation. Therefore, the effective shear stress τ_e can be expressed as:

$$\tau_e = |\tau_n| - \mu\sigma_n \tag{3.4}$$

For the three-dimensional problem, the effective shear stress τ_e initiates frictional sliding. In addition, if τ_e is sufficiently high to overcome the frictional resistance along the initial crack, frictional slip will result in tensile stress concentration at the two tips of the sliding crack, which, in turn, may induce nucleation of "wing cracks". Prior to the onset of crack propagation, the maximum stress intensity factor K_{IImax} of mode II at the initial crack tip is:

$$K_{\mathrm{IImax}} = \frac{4\tau_e\sqrt{a}}{(2-v)\sqrt{\pi}} \tag{3.5}$$

where v is Poisson's ratio, a is statistic radius of disk-shaped crack.

If the wing crack length is infinitesimal, the propagation conditions of disk-shaped crack under triaxial compression based on mixed fracture criterion can be expressed by (Kachanov 1982; Cotterell and Rice 1980):

$$K_{\mathrm{IImax}} \geq K_{\mathrm{IIc}} = \kappa K_{\mathrm{Ic}} \tag{3.6}$$

In Eq. (3.6), K_{Ic} and K_{IIc} are the fracture toughness of mode I and mode II, respectively. κ is proportion factor which may be determined by the mix fracture criterion of linear elastic fracture mechanics. For example, κ is $\sqrt{3(1-v)/(2+2v-v^2)}$ for the maximum-strain-energy criterion for plain stress (Sih and Macdonald 1974; Sih 1974), κ is 0.886 for the maximum-stress criterion (Erdogan and Sih 1963). According to maximum energy release-rate criterion, κ is 1 for maximum energy release-rate criterion (Nuismer 1975).

For the wing crack is caused by the local tensile stress at the disk-shaped crack tip, then the fracture toughness K_{Ic} can be derived by the maximum value of K_{I}:

$$K_{\mathrm{I}} = 2\sigma_t\sqrt{a/\pi} \tag{3.7}$$

where σ_t is local tensile strength at the tip of the disk-shaped crack. Substituting Eqs. (3.5) and (3.7) into Eq. (3.6), then

$$\tau_e \geq \left(1 - \frac{v}{2}\right)\kappa\sigma_t \tag{3.8}$$

Using β to denote $(1-\frac{v}{2})\kappa$, then Eq. (3.8) can be rewritten as:

$$\tau_e \geq \beta\sigma_t \tag{3.9}$$

3.4 The Attitude Angle Parameter φ to Character Micro-fracturing Orientation (Zuo et al. 2015)

Supposing there is a sphere with a unit length radius, and taking the coordinate origin **O** as the center of the sphere. Based on the geometric symmetry of sphere, one of eighth equal parts in the sphere can be chosen as the research object, as shown in Fig. 3.2. There are numerous Griffith micro-cracks distributed randomly in rock materials. Supposing to project one endpoint of any Griffith micro-cracks to

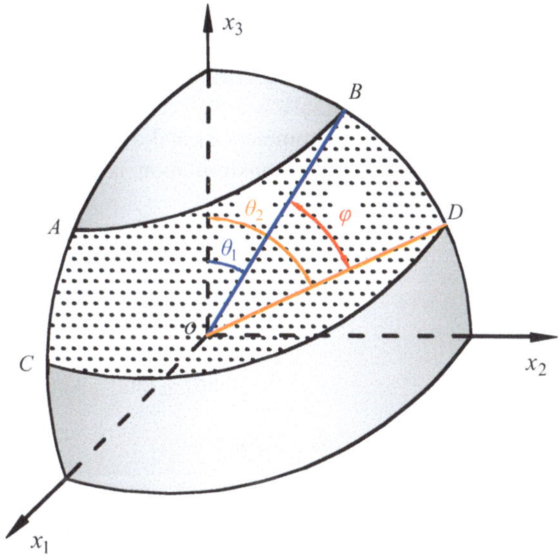

Fig. 3.2 The band region of intersecting point between the normal line of any micro cracks and unit sphere

the coordinate origin O, then there is an intersection point between the normal line of any cracks and the surface of the upper hemisphere. For a rock sample under triaxial compression, all intersection points can make up an intersection region which depends on the three principal stresses. Although evidence has shown that σ_2 does influence the failure of rock (Al-Ajmi and Zimmerman 2005), the shape of the intersection region is rather complicated under true triaxial compression (namely $\sigma_1 > \sigma_2 > \sigma_3$). Therefore, we only discuss the special case of $\sigma_1 > \sigma_2 = \sigma_3$. According to the axial symmetry, these intersection points will make up a band region which locates in two latitude lines AB and CD, and the shadow region is shown in Fig. 3.2.

A sphere including angle θ is used to denote the "micro-fracturing orientation angle" under triaxial compression. The included angle θ is corresponding to the angle α in reference (Zuo et al. 2008), and they have the same physical meaning, namely, "micro-fracturing orientation angle". The difference is that θ is a spatial angle, whereas α is a plane angle. Therefore, when $\sigma_2 = \sigma_3$, Eq. (3.2) can be derived as:

$$\sigma_n = \sigma_1 \cos^2 \theta + \sigma_3 \sin^2 \theta \tag{3.10}$$

The derivation has used Eq. (3.1) and $\cos\theta = l$. Then Eq. (3.3) can also be simplified:

$$\tau_n^2 = l^2(1 - l^2)(\sigma_1 - \sigma_3)^2 \text{ or } \tau_n = \cos\theta \sin\theta(\sigma_1 - \sigma_3) \tag{3.11}$$

From Eqs. (3.4), (3.9), (3.10) and (3.11), one obtains:

$$\cos\theta \sin\theta(\sigma_1 - \sigma_3) - \mu(\cos^2\theta\sigma_1 + \sin^2\theta\sigma_3) \geq \beta\sigma_t \tag{3.12}$$

Let dimensionless parameters $\bar{\sigma}_1$ and $\bar{\sigma}_3$ stand for σ_1/σ_t and σ_3/σ_t, respectively. Corresponding triangular transformation is carried out, then Eq. (3.12) is transformed as:

$$(\mu\bar{\sigma}_3 + \beta)\tan^2\theta - (\bar{\sigma}_1 - \bar{\sigma}_3)\tan\theta + (\mu\bar{\sigma}_3 + \beta) \leq 0 \tag{3.13}$$

Then:

$$\tan\theta_1 \leq \tan\theta \leq \tan\theta_2 \tag{3.14}$$

where,

$$\tan\begin{matrix}\theta_1\\\theta_2\end{matrix} = \frac{(\bar{\sigma}_1 - \bar{\sigma}_3) \mp \sqrt{(\bar{\sigma}_1 - \bar{\sigma}_3)^2 - 4(\mu\bar{\sigma}_1 + \beta)(\mu\bar{\sigma}_3 + \beta)}}{2(\mu\bar{\sigma}_1 + \beta)} \tag{3.15}$$

An attitude angle can be defined, namely, $\varphi = \theta_2 - \theta_1$, then:

$$\tan\varphi = \frac{\sqrt{(\bar{\sigma}_1 - \bar{\sigma}_3)^2 - 4(\mu\bar{\sigma}_1 + \beta)(\mu\bar{\sigma}_3 + \beta)}}{\mu(\bar{\sigma}_1 + \bar{\sigma}_3) + 2\beta} \tag{3.16}$$

Equation (3.16) indicates the relationship between attitude angle φ and external loads. In addition, the range of attitude angle φ is from 0 to $\pi/2$.

3.5 The Failure Characteristic Parameter for Brittle Materials

The usual failure process of rock can be approximately described as follows. There are various pre-existing micro-cracks in rock materials that are potential and of vital significance in their brittle behavior. When the external load achieves a critical value, the pre-existing micro-cracks will propagate, and some new micro-cracks will initiate and propagate. These micro-cracks will tend not to propagate in their own planes but swing into an orientation more nearly parallel to the orientation of major principal stress σ_1 (Diederichs 2003). With the increasing of the external load, some sub-critical micro-cracks will initiate, propagate and tend to be stable which is similar to the former step. When numerous microfractures approach a critical condition, the macroscopic rupture will come into being, and then the rock will be broken. The experimental evidence for the growth of micro-fracturing comes to a large degree from observations on dilatancy, acoustic emission, and change in elastic wave speed in the course of loading (Paterson and Wong 2005; Jaeger et al. 2007). In addition, the microfracture is usually restrained by the confining pressure σ_3, namely, the higher the confining pressure is, the smaller the microfracture is. Therefore, a clear inverse relationship exists between the confining pressure σ_3 and the attitude angle φ of the micro-fracturing orientation angle, namely, the higher σ_3 is, the lower φ is.

It is well known that rock failure depends on external loads and rock property. The failure characteristic parameter of rock should describe the failure processes firstly. Rock will fracture when the characteristic parameter approaches a critical value. In addition, the selected failure characteristic parameter should be a constant.

Though the attitude angle φ of micro-fracturing orientation angle relates with the external load and rock material, it is not a constant. Therefore, the φ cannot be selected as the failure characteristic parameter. To search a failure characteristic parameter suiting for rock material, we made the following further discussions. We change the form of Eq. (3.16), namely,

$$\cos\varphi = \frac{\mu}{\sqrt{1+\mu^2}}\left(1 + 2\frac{\bar{\sigma}_3 + \beta/\mu}{\bar{\sigma}_1 - \bar{\sigma}_3}\right) \tag{3.17}$$

When $\bar{\sigma}_3 = 0$, namely, $\sigma_3 = 0$, then we have:

$$\cos\varphi = \frac{\mu}{\sqrt{1+\mu^2}}\left(1 + 2\frac{\beta/\mu}{\bar{\sigma}_1}\right) \tag{3.18}$$

Equation (3.18) indicates that there is an inverse relation between $\cos\varphi$ and $\bar{\sigma}_1$. The different confining pressure $\bar{\sigma}_{3i}$ (i = 0, 1, 2…n, and $\bar{\sigma}_{30} = 0$) is related to the attitude angle φ_i (i = 0, 1, 2…n) of micro failure orientation angle. In addition, the higher $\bar{\sigma}_{3i}$ is, the lower φ_i is. Therefore, if

$$0 < \bar{\sigma}_{31} < \bar{\sigma}_{32} < \cdots < \bar{\sigma}_{3n} \tag{3.19}$$

then

$$\varphi_0 > \varphi_1 > \varphi_2 > \cdots > \varphi_n \tag{3.20}$$

Due to the range of attitude angle φ_i is between 0 to $\pi/2$, we have

$$\cos\varphi_0 < \cos\varphi_1 < \cos\varphi_2 < \cdots < \cos\varphi_n \tag{3.21}$$

Equations (3.19) and (3.21) clearly indicates that the greater $\bar{\sigma}_3$ is, the greater $\cos\varphi$ is. For an invariable confining pressure, namely the invariable dimensionless parameter $\bar{\sigma}_3$, we can find a relationship between $\cos\varphi$ and the dimensionless parameter $\bar{\sigma}_1$. Therefore, for any different $\bar{\sigma}_3$, the approximate relationship curve of $\cos\varphi$ and $\bar{\sigma}_1$ can be shown in Fig. 3.3. The failure curve denoted by the bold curve ascends gradually with $\bar{\sigma}_1$, as shown in Fig. 3.3.

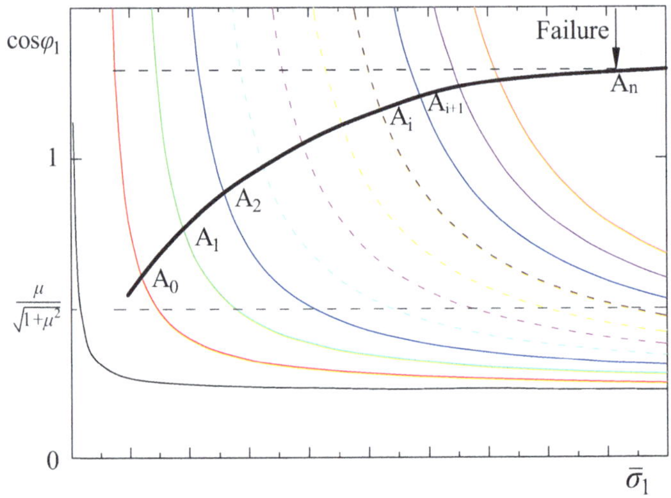

Fig. 3.3 The $\cos\varphi - \bar{\sigma}_1$ curve under different $\bar{\sigma}_3$: the black bold curve is the failure line, the points of intersection, namely, A_0, A_1, …, A_n, are the points of brittle-ductile transition under different confining pressures

3.6 A Theoretical Derivation of the Hoek-Brown Failure Criterion

To discuss the variation of $|\partial\cos\varphi/\partial\bar{\sigma}_1|$ (the change rate of $\cos\varphi$ to axial stress dimensionless parameter) of each point on the failure curve, the following equation can be derived according to Eq. (3.17):

$$\left|\frac{\partial\cos\varphi}{\partial\bar{\sigma}_1}\right| = \frac{\mu}{\sqrt{1+\mu^2}}\frac{2\bar{\sigma}_3 + 2\beta/\mu}{(\bar{\sigma}_1 - \bar{\sigma}_3)^2} \tag{3.22}$$

Transferring the equation form of Eq. (3.17), the following equation can be derived:

$$\frac{1}{(\bar{\sigma}_1 - \bar{\sigma}_3)^2} = [\frac{\sqrt{1+\mu^2}\cos\varphi/\mu - 1}{2\bar{\sigma}_3 + 2\beta/\mu}]^2 \tag{3.23}$$

Substituting Eq. (3.23) into Eq. (3.22), then

$$\left|\frac{\partial\cos\varphi}{\partial\bar{\sigma}_1}\right| = \frac{\mu}{\sqrt{1+\mu^2}}\frac{(\sqrt{1+\mu^2}\cos\varphi/\mu - 1)^2}{2\bar{\sigma}_3 + 2\beta/\mu} \tag{3.24}$$

Any line segment A_iA_{i+1} on failure curve is enlarged, as shown in Fig. 3.4, and the failure point A_i and A_{i+1} correspond to stress state $\bar{\sigma}_{3i}$ and $\bar{\sigma}_{3(i+1)}$, respectively.

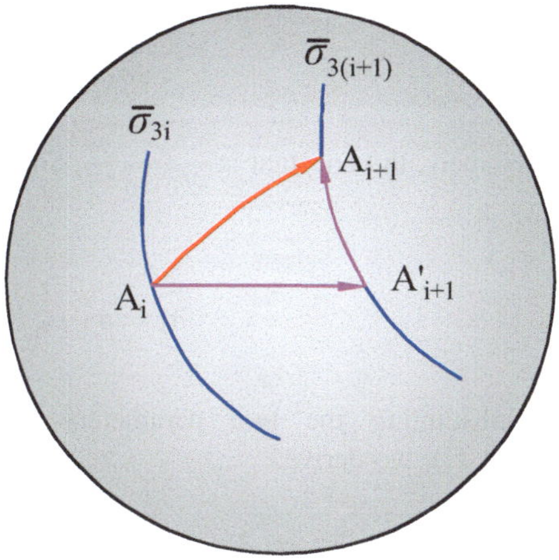

Fig. 3.4 The sketch map of the equivalent relationship

In addition, $\bar{\sigma}_{3i}$ is less than $\bar{\sigma}_{3(i+1)}$. If we only analyze Fig. 3.4 based on geometry, the movement process from A_i to A_{i+1} can be divided into two different courses, namely, first from A_i to A'_{i+1} and then from A'_{i+1} to A_{i+1}. We name them Course $A_iA'_{i+1}$ and Course $A'_{i+1}A_{i+1}$, respectively. In Course $A_iA'_{i+1}$, $\cos\varphi$ remains constant and $\bar{\sigma}_3$ increases; and in Course $A'_{i+1}A_{i+1}$, $\cos\varphi$ increases and $\bar{\sigma}_3$ stays constant.

From Eq. (3.23) and Fig. 3.4, the increase of $\bar{\sigma}_3$ will result in the decrease of $|\partial\cos\varphi/\partial\bar{\sigma}_1|$ when $\cos\varphi$ remains constant in Course $A_iA'_{i+1}$. Contrasting to Course $A_iA'_{i+1}$, the increase of $\cos\varphi$ will result in the increase of $|\partial\cos\varphi/\partial\bar{\sigma}_1|$ when $\bar{\sigma}_3$ stays constant in Course $A'_{i+1}A_{i+1}$. In the two courses, the increment of $|\partial\cos\varphi/\partial\bar{\sigma}_1|$ in Course $A'_{i+1}A_{i+1}$ can compensate for the decrement of $|\partial\cos\varphi/\partial\bar{\sigma}_1|$ in Course $A_iA'_{i+1}$. If the increment of $|\partial\cos\varphi/\partial\bar{\sigma}_1|$ in Course $A'_{i+1}A_{i+1}$ equals to the decrement in Course $A_iA'_{i+1}$, then $|\partial\cos\varphi/\partial\bar{\sigma}_1|$ stays constant. That is to say, any $|\partial\cos\varphi/\partial\bar{\sigma}_1|$ in the failure line is always a constant for any confining pressure when rock fractures. Also, $|\partial\cos\varphi/\partial\bar{\sigma}_1|$ have a simple mathematic expression which can satisfy the former requirements. In other words, $|\partial\cos\varphi/\partial\bar{\sigma}_1|$ can be chosen as the failure characteristic parameter of rock. In addition, $|\partial\cos\varphi/\partial\bar{\sigma}_1|$ should equal to a constant, namely:

$$\left|\frac{\partial\cos\varphi}{\partial\bar{\sigma}_1}\right| = \frac{\mu}{\sqrt{1+\mu^2}}\frac{2\bar{\sigma}_3+2\beta/\mu}{(\bar{\sigma}_1-\bar{\sigma}_3)^2} = \text{constant} \tag{3.25}$$

We can determine the constant through uniaxial compression experiment. Substituting the uniaxial compression conditions $\bar{\sigma}_1 = \sigma_c/\sigma_t = \bar{\sigma}_c$ and $\bar{\sigma}_3 = 0$ into Eq. (3.25), then we have:

$$\left|\frac{\partial\cos\varphi}{\partial\bar{\sigma}_1}\right| = \frac{\mu}{\sqrt{1+\mu^2}}\frac{2\beta}{f\bar{\sigma}_c^2} \tag{3.26}$$

Namely, the constant is $\frac{\mu}{\sqrt{1+\mu^2}}\frac{2\beta}{\mu\bar{\sigma}_c^2}$. Substituting the constant into Eq. (3.25), we derive:

$$\bar{\sigma}_1 = \bar{\sigma}_3 + \sqrt{\frac{\mu}{\beta}\bar{\sigma}_c^2\bar{\sigma}_3 + \bar{\sigma}_c^2} \tag{3.27}$$

Substituting the two parameters $\bar{\sigma}_1 = \sigma_1/\sigma_t$ and $\bar{\sigma}_3 = \sigma_3/\sigma_t$ back into Eq. (3.27), we derive:

$$\sigma_1 = \sigma_3 + \sqrt{\frac{\mu}{\beta}\frac{\sigma_c}{\sigma_t}\sigma_c\sigma_3 + \sigma_c^2} \tag{3.28}$$

where μ is friction coefficient of rock material, which is about 0.2–0.8 (Paterson and Wong 2005); β is a parameter which relates with mix failure criterion; σ_t and σ_c are uniaxial tensile strength and uniaxial compressive strength of rock, respectively. The four parameters depend on rock properties, and all can be obtained from experimental data. σ_3 is confining pressure, and σ_1 is the failure strength corresponding to different confining pressures. This is a theoretical nonlinear failure criterion for rock-like brittle material based on fracture mechanics and numerous existing experimental phenomena, which is similar to Eq. (3.27) in reference (Zuo et al. 2008). The only difference is that the one here is from the three-dimensional condition, the other one in reference (Zuo et al. 2008) is from the two-dimensional condition. The strength criterion also indicates that the failure of rock material is associated with not only the ratio value of uniaxial compression strength σ_c and uniaxial tensile strength σ_t, but also friction coefficient μ and a fracture parameter β which relates to mix fracture criterion. In addition, the failure criterion is not only suitable in the compression zone, but also in the tensile domain. Therefore, it is reasonable to choose $|\partial\cos\varphi/\partial\bar{\sigma}_1|$ as the failure characteristic parameter of rock. Some discussions on the relationship between the criterion and Hoek-Brown criterion can consult the reference (Zuo et al. 2008).

When $\bar{\sigma}_3 = 0$, $\kappa = \sqrt{3}/2$, the $\cos\alpha - \bar{\sigma}_1$ curve changes with μ as shown in Fig. 3.5. When $\bar{\sigma}_3 = 0$, $\mu = 0.3$, the $\cos\alpha - \bar{\sigma}_1$ curve changes with κ as shown in Fig. 3.6.

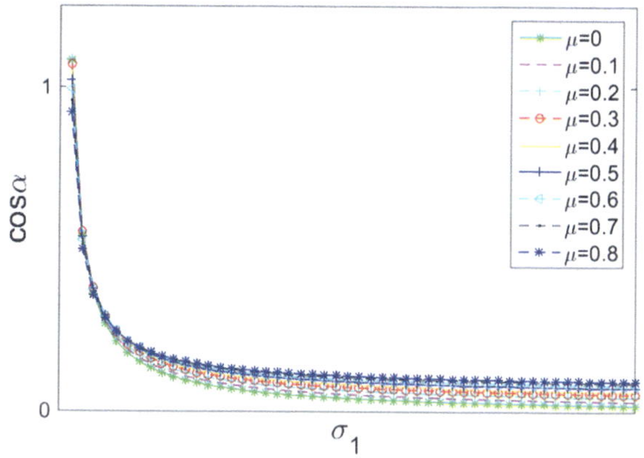

Fig. 3.5 The $\cos\alpha - \bar{\sigma}_1$ curve under different μ

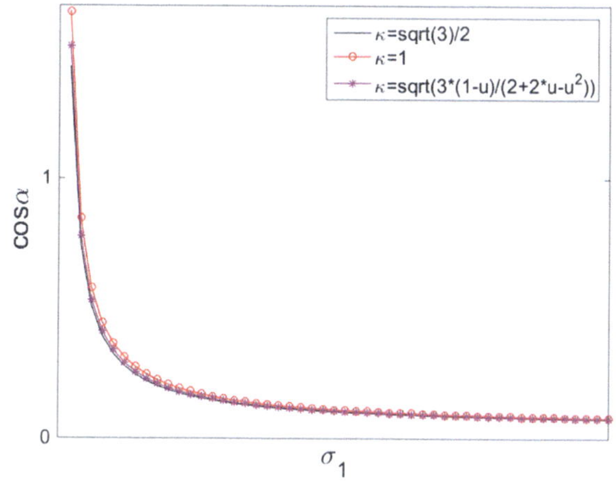

Fig. 3.6 The $\cos\alpha - \bar{\sigma}_1$ curve under different κ

3.7 The Relationship Between Other Trigonometric Functions and Stress

Why can we choose $|\partial\cos\varphi/\partial\bar{\sigma}_1|$ as the failure characteristic parameter of rock? Why not choosing others trigonometric functions? In this section, we will discuss the relationship between other trigonometric functions and stress. And we will explain why other trigonometric functions are not suitable as failure characteristic parameter.

3.7.1 Relationship Between sinα and Stress

First, we discuss $\sin\alpha$. To change the form of Eq. (3.17), namely,

$$\sin\alpha = \frac{\sqrt{(\bar{\sigma}_1 - \bar{\sigma}_3)^2 - 4(\mu\bar{\sigma}_1 + \kappa)(\mu\bar{\sigma}_3 + \kappa)}}{\sqrt{1+\mu^2}(\bar{\sigma}_1 - \bar{\sigma}_3)} \tag{3.29}$$

When $\bar{\sigma}_3 = 0$, we can obtain:

$$\sin\alpha = \frac{\sqrt{1 - 4\kappa\left(\mu\frac{1}{\bar{\sigma}_1} + \frac{\kappa}{\bar{\sigma}_1^2}\right)}}{\sqrt{1+\mu^2}} \tag{3.30}$$

Equation (3.30) shows that there is an inverse relationship between $\sin\alpha$ and $\bar{\sigma}_1$. Therefore, if

$$0 < \bar{\sigma}_{31} < \bar{\sigma}_{32} < \cdots < \bar{\sigma}_{3n} \tag{3.31}$$

then:

$$\alpha_0 > \alpha_1 > \alpha_2 > \cdots > \alpha_n \tag{3.32}$$

Due to the range of attitude angle α_i is between 0 to $\pi/2$, we have

$$\sin\alpha_0 > \sin\alpha_1 > \sin\alpha_2 > \cdots > \sin\alpha_n \tag{3.33}$$

Equations (3.31) and (3.33) clearly indicates that the greater $\bar{\sigma}_3$ is, the smaller $\sin\alpha$ is. Therefore, the parameters of $\sin\alpha$ can neither be selected as the failure characteristic parameters nor reflect the destruction law of the rock. This is one of the reasons.

By finding the partial derivative of Eq. (3.29), we can get:

$$\frac{\partial \sin\alpha}{\partial\bar{\sigma}_3} = \frac{-2(\mu\bar{\sigma}_1 + \kappa)[\mu(\bar{\sigma}_1 + \bar{\sigma}_3) + 2\kappa]\cdot\sqrt{1+\mu^2}}{(1+\mu^2)(\bar{\sigma}_1 - \Sigma_3)^2\sqrt{(\bar{\sigma}_1 - \bar{\sigma}_3)^2 - 4(\mu\bar{\sigma}_1 + \kappa)(\mu\bar{\sigma}_3 + \kappa)}} \tag{3.34}$$

When $\mu = 0.3$, $\kappa = \sqrt{3}/2$, the $\sin\alpha - \bar{\sigma}_1$ curve changes with $\bar{\sigma}_3$ as shown in Fig. 3.7.

Any line segment A_iA_{i+1} on failure curve is enlarged, as shown in Fig. 3.8, and the failure point A_i and A_{i+1} correspond to stress state $\bar{\sigma}_{3i}$ and $\bar{\sigma}_{3(i+1)}$, respectively. In addition, $\bar{\sigma}_{3i}$ is less than $\bar{\sigma}_{3(i+1)}$. If we only analyze Fig. 3.8 based on geometry, the movement process from A_i to A_{i+1} can be divided into two different courses, namely, first from A_i to A'_{i+1} and then from A'_{i+1} to A_{i+1}. We name them Course $A_iA'_{i+1}$ and Course $A'_{i+1}A_{i+1}$, respectively. In Course $A_iA'_{i+1}$, $\sin\alpha$ remains

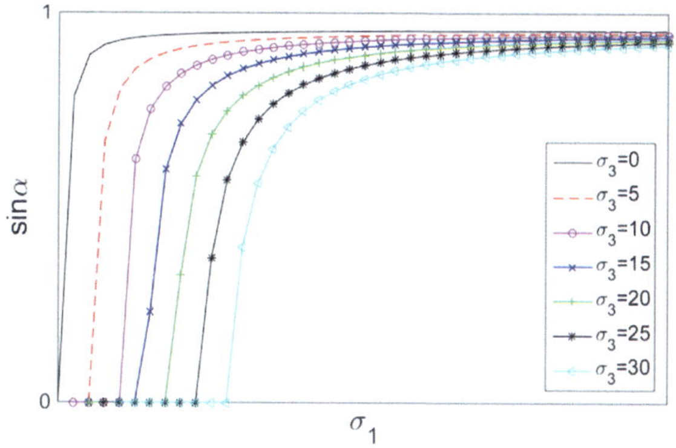

Fig. 3.7 The $\sin\alpha - \bar{\sigma}_1$ curve under different $\bar{\sigma}_3$

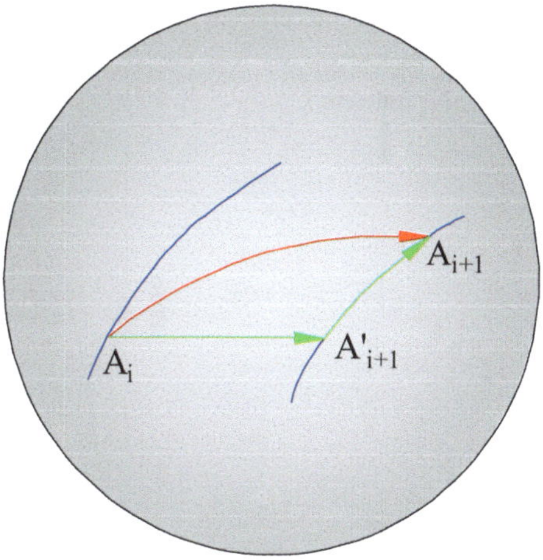

Fig. 3.8 The sketch map of the equivalent relationship

constant and $\bar{\sigma}_3$ increases; and in Course $A'_{i+1}A_{i+1}$, $\sin\alpha$ increases and σ_3 stays constant.

The increase of $\bar{\sigma}_3$ will result in the decrease of $|\partial\sin\alpha/\partial\bar{\sigma}_1|$ when $\sin\alpha$ remains constant in Course $A_iA'_{i+1}$. Contrasting to Course $A_iA'_{i+1}$, the increase of $\sin\alpha$ will also result in the decrease of $|\partial\sin\alpha/\partial\bar{\sigma}_1|$ when $\bar{\sigma}_3$ stays constant in Course $A'_{i+1}A_{i+1}$. In the two courses, $|\partial\sin\alpha/\partial\bar{\sigma}_1|$ is decrease and $|\partial\sin\alpha/\partial\bar{\sigma}_1|$ does not remain constant. Therefore, $|\partial\sin\alpha/\partial\bar{\sigma}_1|$ cannot be selected as a rock failure characteristic parameter.

When $\bar{\sigma}_3 = 0$, $\kappa = \sqrt{3}/2$, the $\sin\alpha - \bar{\sigma}_1$ curve changes with μ as shown in Fig. 3.9.

When $\bar{\sigma}_3 = 0$, $\mu = 0.3$, the $\sin\alpha - \bar{\sigma}_1$ curve changes with κ as shown in Fig. 3.10.

It can be seen from Fig. 3.9 and Fig. 3.10 that the $\sin\alpha - \bar{\sigma}_1$ curve changes with the change of μ, κ, which means that $|\partial\sin\alpha/\partial\bar{\sigma}_1|$ is related to parameters such as μ, κ, and the change of μ, κ causes $|\partial\sin\alpha/\partial\bar{\sigma}_1|$ to change. Therefore, $|\partial\sin\alpha/\partial\bar{\sigma}_1|$ is

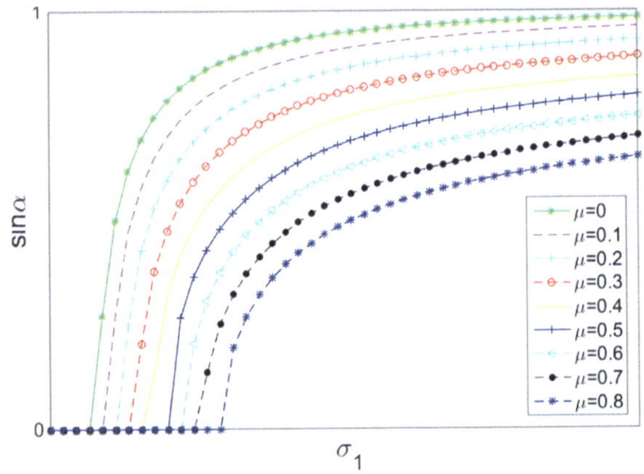

Fig. 3.9 The $\sin\alpha - \bar{\sigma}_1$ curve under different μ

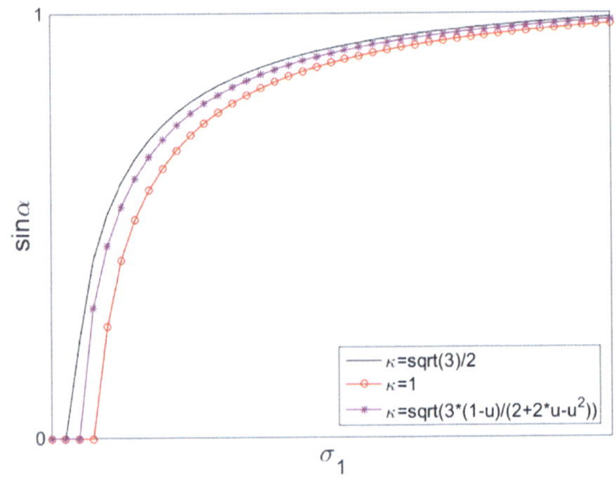

Fig. 3.10 The $\sin\alpha - \bar{\sigma}_1$ curve under different κ

also not a constant. This confirms that $|\partial\sin\alpha/\partial\bar{\sigma}_1|$ cannot be selected as a rock failure characteristic parameter on the other hand.

3.7.2 Relationship Between tanα and Stress

Using Eqs. (3.17) and (3.29), we have:

$$\tan\alpha = \frac{\sqrt{(\bar{\sigma}_1 - \bar{\sigma}_3)^2 - 4(\mu\bar{\sigma}_1 + \kappa)(\mu\bar{\sigma}_3 + \kappa)}}{\mu(\bar{\sigma}_1 + \bar{\sigma}_3) + 2\kappa} \tag{3.35}$$

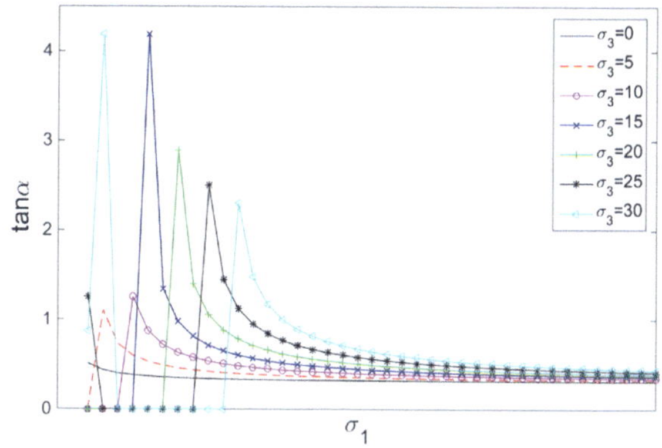

Fig. 3.11 The $\tan\alpha - \bar{\sigma}_1$ curve under different $\bar{\sigma}_3$

When $\bar{\sigma}_3 = 0$, we can get:

$$\tan\alpha = \frac{\sqrt{\bar{\sigma}_1^2 - 4\kappa(\mu\bar{\sigma}_1 + \kappa)}}{\mu\bar{\sigma}_1 + 2\kappa} \tag{3.36}$$

According to Eq. (3.36), there is no linear relationship between $\tan\alpha$ and $\bar{\sigma}_1$, so the parameter about $\tan\alpha$ cannot be selected as the failure characteristic parameter, and cannot reflect the failure law of rock. This is one of the reasons.

When $\mu = 0.3$, $\kappa = \sqrt{3}/2$, the $\tan\alpha - \bar{\sigma}_1$ curve changes with $\bar{\sigma}_3$ as shown in Fig. 3.11.

When $\bar{\sigma}_3 = 0$, $\kappa = \sqrt{3}/2$, the $\tan\alpha - \bar{\sigma}_1$ curve changes with μ as shown in Fig. 3.12.

When $\bar{\sigma}_3 = 0$, $\mu = 0.3$, the $\tan\alpha - \bar{\sigma}_1$ curve changes with κ as shown in Fig. 3.13.

From the above three figures, it can be seen that the $\tan\alpha - \bar{\sigma}_1$ curve is not strictly monotonic and cannot describe the damage characteristics of the rock.

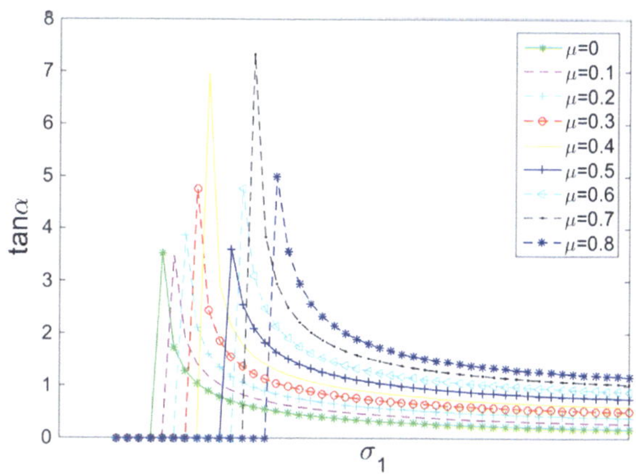

Fig. 3.12 The $\tan\alpha - \sigma_1$ curve under different μ

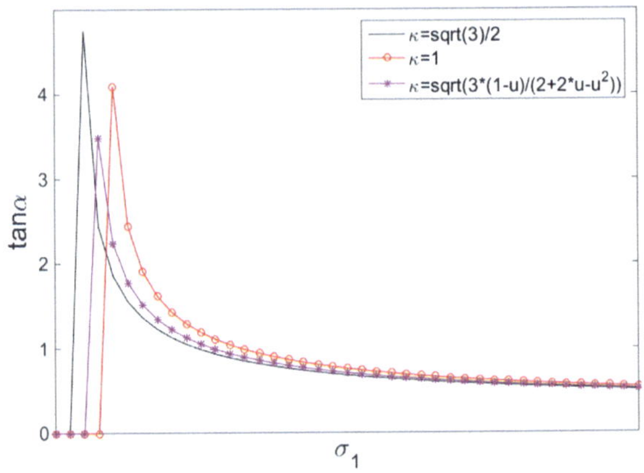

Fig. 3.13 The $\tan\alpha - \bar{\sigma}_1$ curve under different κ

Therefore, $|\partial\tan\alpha/\partial\bar{\sigma}_1|$ cannot be selected as the rock failure characteristic parameter.

3.7.3 Relationship Between cotα and Stress

According to $\cot\alpha = \frac{1}{\tan\alpha}$, substituting the Eq. (3.35) into the above equation, namely,

$$\cot\alpha = \frac{1}{\tan\alpha} = \frac{\mu(\bar{\sigma}_1 + \bar{\sigma}_3) + 2\kappa}{\sqrt{(\bar{\sigma}_1 - \bar{\sigma}_3)^2 - 4(\mu\bar{\sigma}_1 + \kappa)(\mu\bar{\sigma}_3 + \kappa)}} \tag{3.37}$$

Similarly, the trend of cotα is similar to that of tanα. According to Eq. (3.37), there is no linear relationship between cotα and $\bar{\sigma}_1$, so the parameter about cotα can neither be selected as the failure characteristic parameter nor reflect the destruction law of rock. This is one of the reasons.

To find the partial derivative of Eq. (3.37) and get:

$$\frac{\partial\cot\alpha}{\partial\bar{\sigma}_3} = \frac{2(1+\mu^2)(\mu\bar{\sigma}_1 + \kappa)(\bar{\sigma}_1 - \bar{\sigma}_3)}{\left[(\bar{\sigma}_1 - \bar{\sigma}_3)^2 - 4(\mu\bar{\sigma}_1 + \kappa)(\mu\bar{\sigma}_3 + \kappa)\right]^{\frac{5}{2}}} \tag{3.38}$$

When $\mu = 0.3$, $\kappa = \sqrt{3}/2$, the $\cot\alpha - \bar{\sigma}_1$ curve changes with $\bar{\sigma}_3$ as shown in Fig. 3.14:

It can be obtained by analyzing Fig. 3.14 that the $\cot\alpha - \bar{\sigma}_1$ curve has a trend which is substantially same as the $\sin\alpha - \bar{\sigma}_1$ curve. It can be obtained by analyzing the $\sin\alpha - \bar{\sigma}_1$ curve and $\cot\alpha - \bar{\sigma}_1$ curve that both $|\partial\sin\alpha/\partial\bar{\sigma}_1|$ and $|\partial\cot\alpha/\partial\bar{\sigma}_1|$

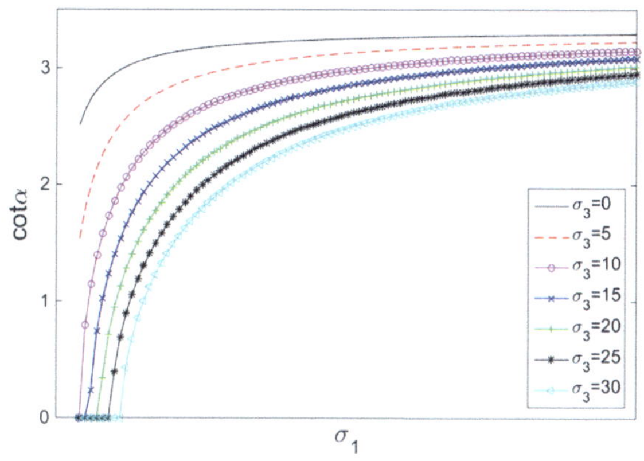

Fig. 3.14 The $\cot\alpha - \bar{\sigma}_1$ curve under different $\bar{\sigma}_3$

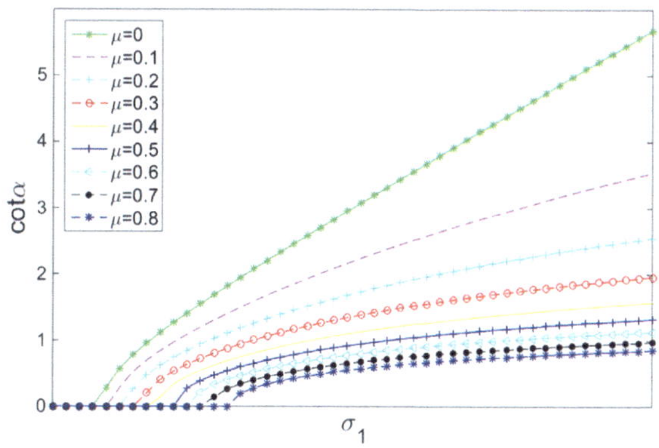

Fig. 3.15 The $\cot\alpha - \bar{\sigma}_1$ curve under different μ

decrease in the processes, and $|\partial\cot\alpha/\partial\bar{\sigma}_1|$ is not a constant. Therefore, $|\partial\cot\alpha/\partial\bar{\sigma}_1|$ cannot be selected as the failure parameter of rock failure feature.

When $\bar{\sigma}_3 = 0$, $\kappa = \sqrt{3}/2$, the $\cot\alpha - \bar{\sigma}_1$ curve changes with μ as shown in Fig. 3.15.

When $\bar{\sigma}_3 = 0$, $\mu = 0.3$, the $\tan\alpha - \bar{\sigma}_1$ curve changes with κ as shown in Fig. 3.16.

By comparing the above changes of $\cos\alpha - \bar{\sigma}_1$ curve, $\sin\alpha - \bar{\sigma}_1$ curve, $\tan\alpha - \bar{\sigma}_1$ curve, and $\cot\alpha - \bar{\sigma}_1$ curve with $\bar{\sigma}_3$, μ and κ, it can be found that: $\cos\alpha - \bar{\sigma}_1$ curve does not change with μ and κ, and the other three curves all change with μ and κ. This indicates that the change of μ and κ does not cause a significant change in the $\cos\alpha - \bar{\sigma}_1$ curve, showing that $|\partial\cos\alpha/\partial\bar{\sigma}_1|$ (the rate of change of the axial stress dimensionless parameter) can be guaranteed to be a constant, not a variable. This also strongly proves the rationality of choosing $|\partial\cos\alpha/\partial\bar{\sigma}_1|$ as a characteristic parameter of rock failure.

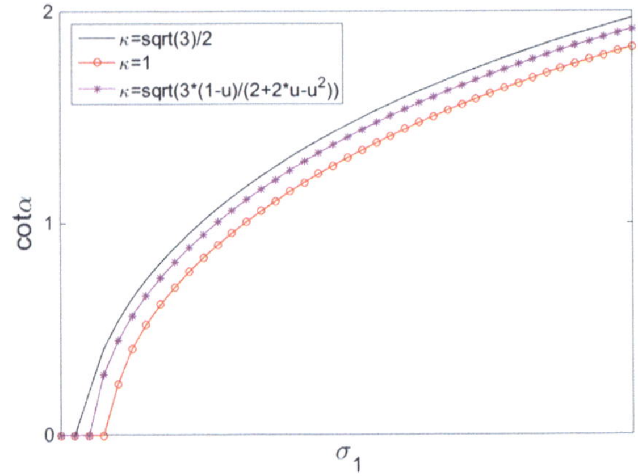

Fig. 3.16 The $\cot\alpha - \bar{\sigma}_1$ curve under different κ

References

Al-Ajmi AM, Zimmerman RW (2005) Relation between the Mogi and the Coulomb failure criteria. Int J Rock Mech Min Sci 42(3):431–439

Anderson TL (2005) Fracture mechanics: fundamentals and applications. Taylor & Francis, Boca Raton, FL

Cotterell B, Rice JR (1980) Slightly curved or kinked cracks. Int J Fract 16(2):155–169

Diederichs MS (2003) Rock fracture and collapse under low confinement conditions. Rock Mech Rock Eng 36(5):339–381

Erdogan F, Sih GC (1963) On the crack extension in plates under plane loading and transverse shear. J Basic Eng 85:519–527

Golshani A, Okui Y, Oda M, Takemura T (2006) A micromechanical model for brittle failure of rock and its relation to crack growth observed in triaxial compression tests of granite. Mech Mater 38:287–303

Griffith AA (1921) The phenomena of rupture and flow in solids. Philos Trans R Soc Lond Ser A 221:163–198

Hoek E, Brown ET (1980) Empirical Strength criterion for rock mass. J Geotech Eng Div 106 (9):1013–1035

Hoek E, Brown ET (1997) Practical estimates of rock mass strength. Int J Rock Mech Min Sci 34:1165–1186

Holcomb DJ (1978) A quantitative model of dilatancy in dry rock and its application to Westerly granite. J Geophys Res 83(B10):4941–4950

Jaeger JC, Cook NGW, Zimmerman RW (2007) Fundamentals of rock mechanics, 4th edn. Blackwell, Oxford

Jian-ping Zuo, Yang Zhao, Neng-bin Chai, Huai-wen WANG (2011) Measuring micro/meso deformation field of geo-materials with SEM and digital image correlation method. Adv Sci Lett 4(4–5):1556–1560

Kachanov ML (1982a) A microcrack model of rock inelasticity, Part I: Frictional sliding on microcracks. Mech Mater 1:19–27

Kachanov ML (1982b) A microcrack model of rock inelasticity, Part II: propagation of microcracks. Mech Mater 1:29–41

Li YP, Yang CH (2006) On sliding crack model for brittle solids. Int J Fract 142(3–4):323−330

Li LC, Yang TH, Liang ZZ (2011) Numerical investigation of groundwater outbursts near faults in underground coal mines. Int J Coal Geol 85(3):276–288

Li LC, Tang CA, Wang SY (2012) A numerical investigation of fracture infilling and spacing in layered rocks subjected to hydro-mechanical loading. Rock Mech Rock Eng 45(5):753–765

Mura T (1987) Micromechanics of defects in solids. Martinus Nijhoff Publishers

Nemat-Nasser S, Hori M (1993) Micromechanics overall properties of heterogeneous materials. North-Holland, Netherlands

Nuismer RJ (1975) An energy release rate criterion for mixed mode fracture. Int J Fract 11(2):245–250

Pan PZ, Feng XT, Hudson JA (2012a) The influence of the intermediate principal stress on rock failure behaviour: a numerical study. Eng Geol 124(1):109–118

Pan PZ, Feng XT, Zhou H (2012b) Development and applications of the elasto-plastic cellular automaton. Acta Mech Solida Sin 25(2):126–143

Paterson MS, Wong TF (2005) Experimental rock deformation-the brittle field. Springer, Berlin

Peng S, Johnson AM (1972) Crack growth and faulting in cylindrical specimens of Chelmsford granite. Int J Rock Mech Min Sci Geomech Abstr 9(1):37–86

Schovanec L (1986) A Griffith crack problem for an inhomogeneous elastic material. Acta Mech 58(1–2):67–80

Sih GC (1974) Strain-energy-density factor applied to mixed mode crack problems. Int J Fract 10 (3):305–321

Sih GC, Macdonald B (1974) Fracture mechanics applied to engineering problems—strain-energy density fracture criterion. Eng Fract Mech 6(2):361–386

Vardoulakis I, Papamichos E (1991) Surface instabilities in elastic anisotropic media with surface-parallel Griffith cracks. Int J Rock Mech Min Sci Geomech Abstr 28(2–3):163–173

Wang XS, Yan CK, Li Y, Xue YB, Meng XK, Wu BS (2012a) SEM in-situ study on failure of nanocrystal metallic thin films and substrate structure under three point bending. Int J Fract 151:269–279

Wang XS, Zhang M, Song XP, Jia S, Kawagoishi N (2012b) Fatigue failure analyses on a Ti-45Al-8Nb-0.2 W-0.2B-0.1Y alloy at different temperatures. Materials 5(11):2280–2291

Yu MH (2004) Unified strength theory and its applications. Springer, Berlin

Zuo Jian-ping XIE, He-ping Zhou Hongwei, Su-ping Peng (2010) SEM in-situ investigation on thermal cracking behavior of Pingdingshan sandstone at elevated temperatures. Geophys J Int 181(2):593–603

Zuo JP, Li HT, Xie HP, Ju Y, Peng SP (2008) A nonlinear strength criterion for rock like materials based on fracture mechanics. Int J Rock Mech Min Sci 45(4):594−599

Zuo Jianping, Liu Huihai, Li Hongtao (2015) A theoretical derivation of the Hoek-Brown failure criterion for rock materials. J Rock Mech Geotech Eng 7(4):361–366

Chapter 4
The Hoek-Brown Brittle-Ductile Transition Analysis

Abstract With the increase of mining depth of coal resources, the mechanical behavior of deep rock is different from that of shallow rock. The mechanical behavior of rock shows the phenomenon of brittle-ductile transition under coupled effects of temperature, pressure and water, which are thought to be the typical factors in deep mine. In this chapter, the mechanism of brittle-ductile transition of deep rock is discussed in detail. It is considered that temperature and pressure plays an external role in brittle-ductile transition, while the change of micro-structure in rock plays an internal role. The brittle-ductile transition of deep rock is affected by two kinds of factors, and they influence each other.

4.1 Introduction

In recent years, with the development of deep coal resources, the mechanical behavior of rocks under high temperature and high pressure has attracted much attention. In deep rock, many rocks tend to show the characteristic of brittle-ductile transition. The study of the brittle-ductile transition behavior of deep rocks not only has the significance of academic research, but also the prospect of far-reaching engineering application. Therefore, it is not surprising that many scholars pay attention to this issue. It can be said that the research on the brittleness, quasi-brittleness and ductile failure of deep rocks under the coupling of temperature and pressure is still not clear, partly due to the complexity of breaking fractures of rock materials composed of various crystals. Part of the damage caused by the internal occurrence of the Earth is often masked by other geological processes. Some gratifying results have been achieved (Handin 1953; Paterson 1958; Heard 1960; Jaeger et al. 2007; Mogi 1965, 1966; Gowd and Rummel 1980; Sibson 1977, 1980; Meissner and Kusznir 1987; Ranalli and Murphy 1987; Kwasniewski 1989; Shimada 1993; Paterson 1982) by many scholars, especially the experimental phenomena have been discussed a lot, but the discussion on the mechanism of meso-deformation and damage is rarely reported. In this paper, the experimental study on brittleness ductility transition of rock is summarized and the influence of the microstructure of rock minerals on the brittle-ductile transition is discussed.

J. Zuo and J. Shen, *The Hoek-Brown Failure criterion—From theory to application*,
https://doi.org/10.1007/978-981-15-1769-3_4

4.2 Experimental Researches on the Brittle-Ductile Transition of Rock

It is generally believed that the material is brittle without or with a small amount of permanent deformation. In contrast, a material with permanent deformation when broken is said to be ductile. In mining, with the increase of mining depth, the mechanical properties of rock will change obviously. One of the characteristics of this change is the transformation of the rock failure mechanism; it is from the brittle failure in the shallow part to the ductile failure in the deep mining conditions. To be precise, it is actually the dynamic damage behavior from the shallow part to the deep quasi-static destruction behavior. The mechanical behavior of rocks in the brittle-ductile transition region is obviously restricted by strain rate, temperature, effective stress, the microstructure of water and minerals. The brittle deformation and crystal plasticity are not independent and the strength of rock is affected by various deformation mechanisms. The determination of the transition point (position) of brittle and ductile deformation is one of the key problems in the study of rheological properties varying with depth, which is generally considered to be of important dynamic significance.

Since Von Karman's first experiment with Carrara marble, many studies have been conducted on the effect of confining pressure on rock ductility (e.g. the effect of pore fluid pressure on effective confining pressure). Handin (1953) observed a brittle-ductile transition of rock salt under the condition that the confining pressure is less than 20Mpa in the triaxial compression test at room temperature. Figure 4.1 shows the failure of Wombeyan (Paterson 1958) marble under different confining pressures on a servo-controlled test machine, demonstrating the characteristics of rock transition from brittle to ductile with increasing pressure. It can be seen from Fig. 4.1 that under higher confining pressure, the stress level will increase until failure, and we can see that the increase of confining pressure will make the load-displacement curve after rock rupture tends to be gentle. The peak stress

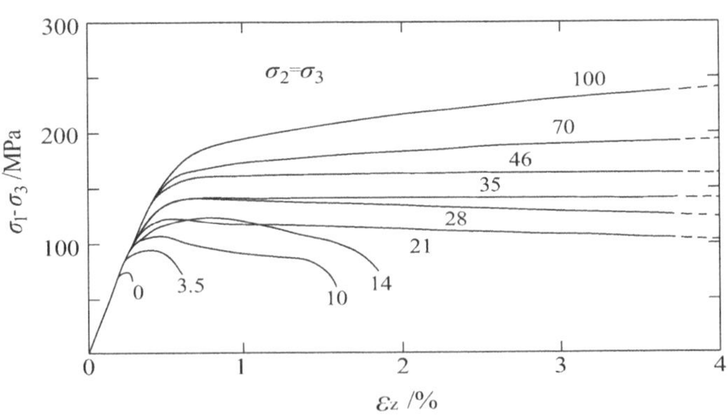

Fig. 4.1 Stress-strain curve of the triaxial compression test of Wombeyan marble (Paterson 1958)

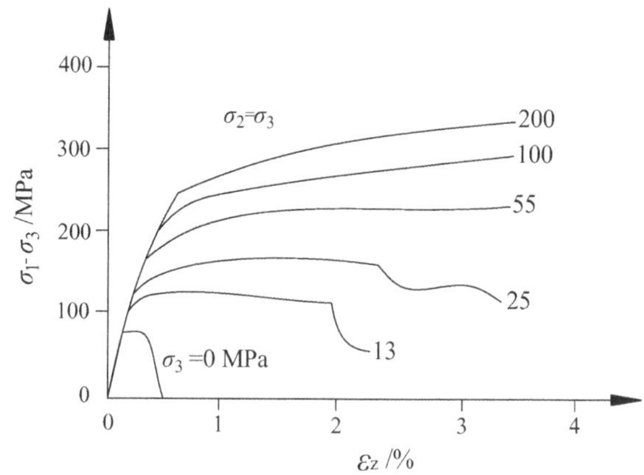

Fig. 4.2 Experimental results of different pressures of Yamaguchi marble (Mogi 1965)

occurs at a larger strain and a rough critical confining pressure can be seen. When the confining pressure is lower than the critical value, the bearing capacity of the rock decreases with the increase of the strain and the strain is relatively small at the time of the failure. At this time, the rock exhibits brittle properties. When the confining pressure is higher than the critical value, the rock cannot lose its bearing capacity under the larger strain, and the bearing capacity will even be improved, then the rock shows ductility. It indicates that when the confining pressure increases to a critical confining pressure of bridging transition, the rock will undergo a brittle to ductile transition. Heard (1960) used a quantitative explanation for the brittle-ductile transition, and the strain value at the time of failure was 3–5%, which was called the beginning of the brittle-ductile transition. Jaeger et al. (2007) took a qualitative view that the rock is in a ductile stage as long as it can withstand permanent deformation without losing its bearing capacity. If the rock has a lower bearing capacity as the deformation increases, it is said that the rock is in the brittle phase.

In the following years, Mogi (1965, 1966) also found similar results through experiments, and found that brittle-ductile transitions are usually related to rock strength. As shown in Fig. 4.2, when the confining pressure is approximately equal to 1/3 of the principal stress difference $(\sigma_1 - \sigma_3)$ at the time of failure, the silicate rock undergoes a brittle-ductile transition. While for the carbonate rock, the confining pressure is slightly smaller, and more pressure is required to make the rock ductile in the tensile test (Fig. 4.2).

The effect of confining pressure on rock deformation varies with different rock samples. Figure 4.3 is the effect of confining pressure on the deformation of porous sandstone by increasing the confining pressure to 200 MPa (Gowd and Rummel 1980). It can be seen that the axial stress is less than the yield strength, the deformation of the sandstone is basically linear elastic, and the magnitude of the yield strength obviously depends on the confining pressure. When the confining pressure is relatively low, such as confining pressure $\sigma_3 < 90$ MPa, the stress will decrease with the increase of strain after the stress reaches the peak stress, and will

Fig. 4.3 Stress-strain relationship of Bunt sandstone under different confining pressures

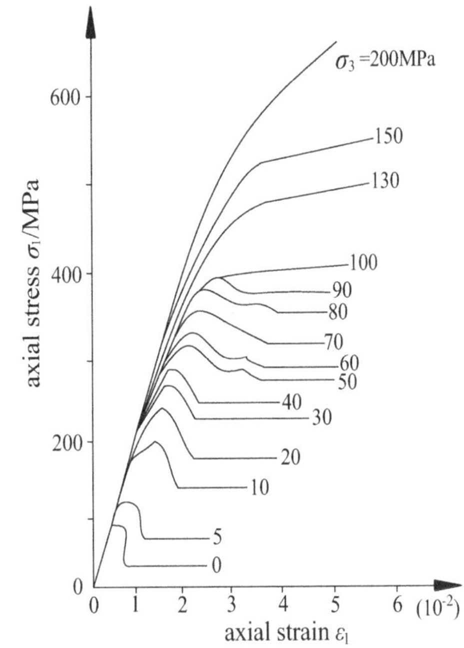

remain stable when the stress is reduced to a residual strength. The stress will keep steady as the deformation increases. When the confining pressure exceeds 100 MPa, the sandstone exhibits hardening properties.

Some scholars have also explained the brittleness ductility transition by other methods. Sibson (1977, 1980) has noticed that there is a high-stress release in the depth of the transition from brittleness to ductile flow. Meissner and Kusznir (1987) and Ranalli and Murphy (1987) respectively proposed that the frictional strength is compared with the creep strength, and the equivalence is the depth from brittleness to ductility. After summarizing a large amount of experimental data, Kwasniewski (1989) believed that there is also an intermediate transition state in brittle-ductile transformation. According to the experimental results of the high-pressure fracture type, Shimada (1993) believes that the rheological properties of the rock layer at a dozen kilometers are controlled by brittle fracture, and the fracture strength and creep strength are equal to the brittle-ductile conversion depth. In addition, some rocks still exhibit brittleness even under high confining pressure at 1000 MPa (Paterson 1982).

During the brittle-ductile transition of rock, the effect of temperature is also great. Generally, as the temperature increases, the strength of the rock will decrease, but the thermal activation of the rock mineral particles and the influence of the temperature gradient will make the rock more prone to arise brittle-ductile transition.

Although a large number of experiments show that the damage mechanism of rock changes from brittleness to ductility with the increase of confining pressure and temperature, these macroscopic qualitative explanations are very helpful for us to understand the damage mechanism of deep rocks. But it is known that any mechanical properties of a solid are ultimately determined by the size of the atom or molecular scale and the size of the micro-defect. These problems are well reflected in the fields of metal, ceramics and composite materials, but the damage to rock minerals has not been comprehensively studied. The following is a preliminary study on how microstructure affects the brittle-ductile transformation of rock.

4.3 Microstructure Effects of Rock Minerals

Figures 4.1–4.3 are stress-strain curves for a typical brittle-ductile transition. When the confining pressure rises, brittle to ductile transition will begin and many of these mechanisms will start to work. When the confining pressure decreases, the mechanism of brittle fracture will dominate the rock. At present, the transition mechanism of brittleness-ductility is somewhat in the understanding of macroscopic phenomenology, and the mesoscopic mechanism of its transformation is less discussed. Especially, the research on the brittle-ductile transition of the microstructure of rock minerals is rarely reported. And these microstructures have a limitation of further understanding the nature of the rock.

The physical properties of rock minerals should be related to their own microstructures, such as the crystallites of rock minerals and the structure of their crystal lattices, composition, particle size, cementation and crystallization. With the occurrence of deformation, the microstructure of rock minerals has received a lot of attention both in theory and in the experiment. More experimental attention is paid to the effects of particle shape and particle size, particle growth, particle rotation, particle rearrangement, and dislocation behavior.

Rock minerals are composed of a variety of crystals. The size of these crystal particles may vary. However, the structural features of atoms or molecules inside the crystal that are repeatedly arranged in a periodic pattern are common (strict crystal structures are periodic and symmetrical).

All minerals are also characterized by a combination of symmetrical elements. Like other crystals, rock minerals can also be attributed to the following seven crystal systems: triclinic, monoclinic, orthorhombic, trigonal, tetragonal, hexagonal, and equiaxed systems (Nicolas and Poirier 1976). The crystal structure has an effect on the physical and mechanical properties of the rock. Since the elasticity of rock and mineral is first related to the fracture of bonds between atoms and molecules in the internal crystal system of mineral, the change of mineral structure, mineral chemical composition and density under certain thermodynamic conditions can be studied on the basis of the study of elasticity. The inelastic nature of rock is largely determined by the crystal structure of rock mineral crystals and the actual composition of mineral particles, which is partly caused by the plastic deformation result from the slip of atomic and molecular bonds under shear stress. For rock-based materials, the intergranular bond strength between grains is lower than the strength of the grains themselves. Moreover, there are many brittle phases or enrichment of impurity atoms along the grain boundary to make the grain boundary brittleness, and the slip coefficient of the material itself is insufficient. Many crystals inside the rock fail to guarantee microscopic continuity conditions in the deformation, and these factors can cause changes in the rock failure mechanism. For example, magmatic rocks can be classified into an all-crystalline structure, a semi-crystalline structure, and a vitreous structure according to their crystal formation under different conditions. The all-crystalline structure is composed entirely of crystals of minerals. Under confining pressure and high temperature, intergranular, transgranular or intergranular and transgranular brittle fracture modes are usually observed. For magmatic rocks with semi-crystalline or vitreous structures, it is easier to exhibit ductile properties under the action of high confining pressure and high temperature. Therefore, the crystal structure and true organization of these rock minerals and the slip of their internal atoms are a major mechanism for the brittle-ductile transition. Under the action of confining pressure and temperature, the sliding mechanism is dominant because the continued expansion of micro-cracks is inhibited, even closed or healed.

On a macroscopic scale, rock is composed of a variety of mineral components, as well as cement and heterogeneity, and it is a complex geological material. All rocks can also be considered to be composed of a variety of mineral crystals from a microscopic scale. On a microscopic scale, rocks are composed of an infinite number of atoms. These atoms are bonded by the interaction of electromagnetic fields to form mineral particles, and mineral particles are formed by the cementation and bond force of the cemented substance or hetero group. When the rock begins to deform under a certain external load, the elasticity is directly related to the relative motion of the atom. When the bond between these atoms is destroyed, the rock begins to damage and destroy. However, destroying these bond forces requires more energy, and the cement inside the rock is more likely to be deformed and destroyed before the bond is broken. Johnson (Johnson et al. 1978) used accurate acoustic emission techniques to study the thermal cracking of rocks at uniform slow heating rates, pointing out that 75 and 200 °C are the thermal cracking temperatures of granite and quartzite, respectively. Xu (1998), Xu and Liu 2000) studied the Three Gorges granite and reached similar conclusions. It is believed that 75 and 200 °C are the threshold temperatures of granite elastic modulus and uniaxial compressive strength, respectively. However, this temperature range is much lower than the melting temperature of rock mineral crystals. Therefore, it can be considered that the main factors causing the change of rock mineral elastic modulus and compressive strength, when the temperature is below 200 °C, may be the influence of cement and hetero groups. Common cements are siliceous, calcium, iron and mud. Some experimental results (Meng et al. 2000) also indicate that the cement has an effect on the strength of the rock. The strength of rock with siliceous and iron cementation is higher than that of rock with calcareous and muddy cementation. Since the strength of cement or hetero-base is usually much lower than that of mineral crystal particles, mineral crystal particles deform together with cement and hetero-base to reach the elastic limit under the action of the external load. Then, the cement or the hetero group begins to undergo large deformation until the strength limit damage is reached, and the mineral crystal particles begin to plastically deform at this time and some even do not destroy, exhibiting ductile properties. The mode of deformation is also affected by the structure and type of cement or hetero-base, as well as the way of cementation and support.

Most of the rock minerals in the earth contain more or less water or solution, and some places are rich in oil and natural gas. Many scholars believe that rock mass can be treated as a two-phase medium, which is composed of solid phase material (mineral particles) and liquid phase material containing water in pores and fractures (Müller 1981). More strictly speaking, it should also include the gas phase in the gap (such as air, natural gas, etc.). The effect of water on the mechanical properties of rock can be discussed in two aspects: the mechanical effect of water and chemical corrosion. Pure water is neutral and can generally be considered as not corrosive to rock minerals, but most rock minerals contain certain substances that make the water acidic or alkaline, which has a great weakening effect on soluble rocks. For example, for silicate glass and quartz, the general expression of weakening in a water environment is as follows (Atkinson 1992):

$$H-O-H \not\angle [\equiv \mathrm{Si}-O-\mathrm{Si}\equiv] \Leftrightarrow [\equiv \mathrm{Si}-\mathrm{OH}\bullet \mathrm{HO}-\mathrm{Si}\equiv] \Leftrightarrow 2[\equiv \mathrm{Si}-\mathrm{OH}]$$

It can be seen that the strong Si–O bond is hydrolyzed into a weak hydroxyl group bonded to the silicon atom. The stress corrosion of the Si–O–Si bond is attributed to water, and more specifically to the ionized water (Wiederhom 1978) and molecular water (Michalske and Freiman 1982), and both the molecular and ionic water activities play an important role in controlling the stress corrosion rate (Freiman 1984). It also affects the failure mechanism of rock minerals. The mechanical effect of water in rock minerals mainly means that some molecular motion is activated by the addition of water, and the pore pressure is generated by the liquid or gas in the pores and cracks of the rock. Under the action of external load and temperature, the pore pressure will increase, and the pores will extend. On the one hand, the pressure will offset some of the total stress acting on any part of the rock (including the stress caused by confining pressure and tectonic movement), on the other hand, it will weaken the bonding force and cohesive force between the mineral particles and reduce the shear resistance of the rock. In this case, the elastic yield limit of the rock is reduced, and the deformation mechanism is transformed from brittle failure to ductile failure. Therefore, water is also a factor that cannot affect the brittle-ductile transformation of rock. Certainly, the degree of influence is also affected by factors such as water content, rock mineral composition, cementation status, and crystallinity. Under the influence of temperature and pressure coupling, the impact is greater and the rock is more prone to arise brittle-ductile transitions.

Many scholars have studied the relationship between grain size and strength, and derived the following two relations (Li et al. 1988):

$$\sigma = \sigma_{\infty} + \sigma_1 D^{-\frac{1}{2}} \tag{4.1}$$

$$\sigma = \sigma_1 D^{-a} \tag{4.2}$$

In the formula, the σ is the strength, and *D* is the grain diameter, which is the material constant (the fracture of the brittle solid is somewhat data). Carniglia (1972) studied the relationship between grain size and strength of ceramics, and the experimental results are in good agreement with Eq. (4.1). If Eq. (4.1) is slightly changed, an expression similar to Eq. (4.2) can be obtained by substituting σ' for $(\sigma - \sigma_{\infty})$. It can be seen that both of the equations indicate that the strength is exponentially related to the grain size, and the strength increases with the grain size. There is a downward trend. In general, the failure stress of non-cubic brittle ceramics is always increasing when the grain size increases to a certain critical size. When the grain size is further increased, the failure stress is reduced. It is mainly because micro-cracked ceramics develop micro-cracks with many grain boundaries under stress, which absorb some energy that can promote crack propagation. As the grain size increases, these cracks along grain boundaries become less and are not independent, making them easier to bridge and promote crack propagation (Rice and Freiman 1981).

As early as 1921, Griffith (1921) studied the crack propagation problem in quasi-brittle materials such as glass and ceramics, and proposed the energy criterion for crack propagation of brittle materials. It is believed that when the energy released is greater than that required to form new surface, the crack will expand. Certainly, we know that the view is incomplete because it does not consider the energy dissipated by the plastic deformation that occurs before the crack propagates, and in fact this part of the plastic deformation can account for a large part of the specific gravity (Irwin 1957). But Griffith's idea of the principle of energy balance has had a profound impact, and the destruction of materials is thought to be related to the newly formed surface. According to the previous formula, we can analyze it by the energy balance principle of Griffith. The smaller the grain size is, the greater the new surface area is formed when the grain is broken, the greater the fracture surface energy is required when breaking, and therefore the higher the failure intensity will be. And the intensity of the damage is also higher. The fracture energy of the grain boundary is generally smaller than the fracture energy of the grain splitting because impurities, pores, particles of the second phase, and defective bonds at grain boundaries tend to weaken them. In addition, the boundary rupture energy of large grains is often smaller than that of small grains because the density of defects on large grain boundaries is greater. As the depth of mining increases, both temperature and pressure increase. Under their coupling effect, mineral composition, mineral cement and crystal structure (crystal system) have an impact on the final fracture failure of rock minerals. The characteristics of the brittle-ductile transition are obtained, and the brittle-ductile transition is more likely to occur in the minerals of small particles, which is mainly caused by two thermal stress driving forces. One is due to the reduction in surface energy, and the second is due to the reduction in storage strain energy owing to defects or dislocation density in mineral crystal deformation. As the temperature increases, the surface energy of the mineral particles will decrease, which may cause the mineral grains to grow, and the reduction of the strain energy is related to the strain rate of the external load. Nicholson (1972) reached the same conclusion in the study of superplasticity of metals. Experimental observations also show that the growth rate of the particles is high when the initial particle size is small due to the reduction in surface energy. When a new crack surface appears, the crack surface will also heal due to the action of confining pressure and temperature (Lawn 1993). Similar phenomena are found in rock salt (Hou 2003) and mica (Obrimoff 1930). The rate at which new surfaces form and heal, or more precisely the relative rate at which bonds break and join, plays an important role in the brittle-ductile transition of rocks. It is believed that the deformation characteristics of the rock are more likely to exhibit ductile characteristics when the breaking speed of the key is slower than the joining speed of the key. It can be said that the relationship between the strength of the rock and the particle size has an obvious scale effect, and there are obvious scale effects in concrete and other quasi-brittle materials (Watts et al. 1976), but our discussion here is only suitable for a certain scale range. If the particle size is further reduced or increased, how the strength of the rock changes will still be verified by experiments.

The above mainly discusses the influence of the microstructure on the mineral crystal system, the crystal grain size, and the microstructure of the cement. Certainly, other factors, such as the shape of mineral crystals, the migration and slip of grain boundaries, the rotation of grains, and the rearrangement of grains after deformation, all have an effect on the destruction of rock minerals. When the structure of rock minerals is angular, the mineral crystals will mesh with each other and being combined with geometric factors, cement and hetero groups. Under high pressure and temperature in deep mining, due to the geometry of these minerals and the influence of cement and heterogeneity, the rock is more prone to arise brittle-ductile transition.

The coupling effect of temperature and pressure of deep rock mineral crystals is usually accompanied by sliding and growth of grain boundary at the mineral crystal boundary, while migration and growth of grain boundary will cause reduction of grain boundary energy or surface energy, which is also the cause of brittleness ductility transition of rock. Inglis (1913) observed a large number of particle rotations in metal superplastic rheology, but the particle rotation usually does not exceed 45°, and an interesting result is obtained. As the load increases, during the plastic deformation of the metal, the rotation of the particles will constantly adjust the direction of the transformation rotation. In the case of a microscopic layer facing rock mineral crystals, there is also a phenomenon of grain rotation. Studies by Paterson and Wong (2005) and Griffith (1921) have shown that even microscopic defects can be a potential cause of solid weakening. Regardless of whether these defects exist in advance or due to external factors such as external load and temperature, we can predict that these microscopic defects also provide more geometric space for the rotational deformation of the grains. Certainly, under the high temperature and high pressure, it also result in the activation of crystal particles and other surrounding media, and is more conducive to the translation or rotation of the crystal particles. It is generally believed that these crystal particles are very strong. In this case, the crystal particles can be regarded as rigid bodies, so the translation or rotation of the crystal particles does not have a great influence on the entire volume change, but makes the brittle-ductile transition provide some room for deformation, causing the deformation easier to occur freely. In the deformation of rock under the high temperature and high pressure, the translation or rotation of crystal particles, particle slip and growth will lead to the change of the whole crystal structure. At this time, the crystal structure is usually more prone to occur ductile deformation than the original crystal structure.

4.4 Brittle-Ductile Transition Analysis Based on H-B Criterion

It is usually accepted that material failure is a brittle failure if no apparent plastic deformation takes place before fracture. In ductile failure, by contrast, extensive plastic deformation takes place before fracture. Experimental results indicate that although most of the rocks exhibit brittleness under atmospheric pressure, ductility in rock can also be achieved in the laboratory with the aid of sufficiently high confining pressure and temperature (Jaeger et al. 2007; Heard 1960). In other words, there exists a critical condition for the transition from brittleness to ductility. In this paper, we are concerned about the effects of confining pressure but not the effects of temperature. The brittle-ductile transition will occur when external confining pressure achieves a critical value. The state where this change happens is called the brittle-ductile transition state. In addition, the investigation on brittle-ductile transition has become of central importance in many geologic situations but which may also have relevance to some engineering contexts.

Since the first works of von Kármán (1911), there are many experimental studies which have been reported on the role of confining pressure in achieving ductility (Mogi 1965, 1966). However, most of them are from macroscopic view or phenomenology, taking no account of the micro-mechanisms when the deformation occurs, especially not including the micro-cracks mechanism under high confining pressure. For example, some literatures artificially defined that the failure of rock is a brittle failure if the value of strain or confining pressure of failure is less than a specific value, by contrast, ductile failure occur when the value is greater than a specific value. For example, the value of 3–5% strain to failure is often taken to define the brittle-ductile transition (Evans et al. 1990; Wong et al. 1997). Some authors proposed that, as a general rule, the brittle-ductile transition should be related to the strength of rocks (Mogi 1965, 1966). For instance, in silicate rocks in compression, it occurs when the confining pressure becomes equal to roughly 1/3 of the stress difference at failure, and to 1/4 in carbonate rocks. Ductility of rock in extension requires much higher confining pressures (Handin 1966). However, in the case of porous sandstone, the transition to ductility occurs at an effective confining pressure of about 0.15 times than the critical effective pressure for the onset of grain crushing under hydrostatic pressure (Handin 1966). In other words, there is not a strict or unified standard to choose a critical parameter of the brittle-ductile transition. The parameter selection varies with individuals or rock types. This shows that it is difficult to make clear and strict physical standards of the brittle-ductile transition point at a macro scale study of rock. Therefore, it is necessary to carry out deep insights on the concepts of brittle, ductile and brittle-ductile transition point at a micro/meso scale. This section in the present work shows some new discussions on the subject.

Ductility has been defined by Handin (1966) as the ability to undergo large permanent deformation without fracture (Brace et al. 1966). With the increase of confining pressure, ductility increases markedly and a transition from the brittle to the ductile state takes place at some confining pressures. In light of Handin's suggestion and our former failure criterion, we define that the brittle-ductile transition occurs in rock when the confining pressure achieves a suitable value. And the value is achieved when the micro-fracturing in rock is completely restrained, including both microcracks initiation and propagation. This is a definition of rock brittle-ductile transition point at a micro/meso scale. Since there is a clear relationship between micro-fracturing and rock dilatancy, this definition is entirely equivalent to the macro definition (Mogi et al. 2007). According to the definition, the confining pressure value of brittle-ductile transition point can be derived as follows. In the former part of this paper, an attitude angle φ has been used to characterize microfracturing orientation, and the relationship between the angle φ and external load has also been derived. It has also been proved that selecting $|\partial\cos\varphi/\partial\bar{\sigma}_1|$ as rock brittle failure characteristic parameter can not only meet experimental results, namely, the larger the confining pressure is, the smaller the micro-fracturing density is, but the results derived are in accordance with the Hoek-Brown empirical criterion.

Obviously, when the confining pressure increases to a specific value $\bar{\sigma}_3^*$, under which the internal micro-fracturing is completely restrained when the rock is broken, the corresponding value of φ is 0. This corresponds to the intersection of the bold failure line ($|\partial\cos\phi/\partial\bar{\sigma}_1| = \text{constant}$) and horizontal line ($\cos\varphi = 1$), as shown in Fig. 4.3. According to Eqs. (3.24) and (3.26), the equation of failure line can be written as:

$$\left|\frac{\partial\cos\varphi}{\partial\bar{\sigma}_1}\right| = \frac{\mu}{\sqrt{1+\mu^2}}\frac{(\sqrt{1+\mu^2}\cos\varphi/\mu - 1)^2}{2\bar{\sigma}_3 + 2\beta/\mu} = \frac{\mu}{\sqrt{1+\mu^2}}\frac{2\beta}{\mu\bar{\sigma}_c^2} \tag{4.3}$$

Substituting $\varphi = 0$ and $\bar{\sigma}_3 = \bar{\sigma}_3^*$ into Eq. (4.3), we can obtain:

$$\bar{\sigma}_3^* = \frac{1}{\mu}[\frac{\bar{\sigma}_c^2}{4\beta}(\sqrt{1+\mu^2} - \mu)^2 - \beta] \tag{4.4}$$

Equation (4.4) is the critical transition condition of brittle-ductile transition for rocks. It indicates that the greater $\bar{\sigma}_c = \frac{\sigma_c}{\sigma_t}$ of rock is, the greater the confining pressure $\bar{\sigma}_3^*$ that triggers brittle-ductile transition is. This is consistent with the experimental results. In addition, the critical confining pressure of brittle-ductile transition is also related with the friction coefficient μ and a fracture parameter β of rocks.

An empirical failure criterion has also been proposed by Mogi (1966), namely, for most rocks, the confining pressure must always smaller than the uniaxial compressive strength to keep brittle behavior of rock:

$$\bar{\sigma}_3^* \leq \bar{\sigma}_c \tag{4.5}$$

To compare the two criterions of Eqs. (4.4) and (4.5), we plot them in Fig. 4.4. In Fig. 4.4, considering brittle-ductile behavior in the conventional triaxial compression test as a function of the confining pressure and compressive strength of Silicate rock and Carbonate rock were also plotted (Zuo et al. 2015). The transition by Mogi (1966) is linear.

However, most experimental data in Fig. 4.4 shows that the brittle-ductile transition relationship may be nonlinear. Equation (4.4) can approximately agree well with experimental data for low compressive strength rock. The figure also shows that, only when $2\beta(\sqrt{1+\mu^2}+\mu) \leq \bar{\sigma}_c \leq \frac{2\beta}{\sqrt{\mu^2+(\sqrt{1+\mu^2}-\mu)^2}-\mu}$, Eq. (4.5)

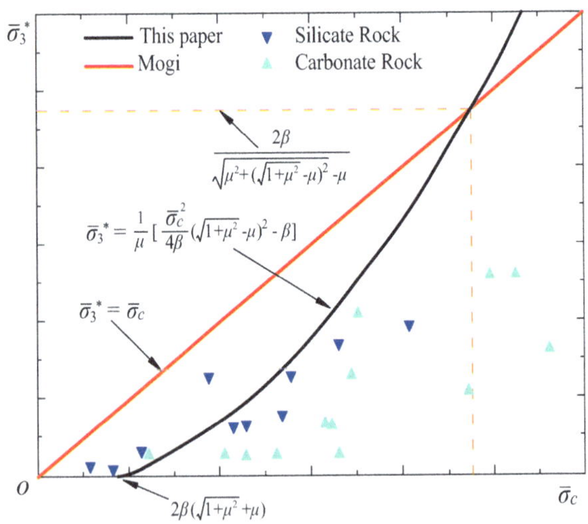

Fig. 4.4 The relationship between the confining pressure $\bar{\sigma}_3^*$ at brittleness-ductility transition and the value of $\bar{\sigma}_c$

comes into existence. This also means that Eq. (4.4) is in accord with Mogi's criterion under a certain condition. Figure 4.4 also shows that when the value of $\bar{\sigma}_c$ is less than $2\beta(\sqrt{1+\mu^2}+\mu)$, rock does not exhibit brittle behavior under compressive compression. In addition, when $\bar{\sigma}_c$ is more than $\frac{2\beta}{\sqrt{\mu^2+(\sqrt{1+\mu^2}-\mu)^2}-\mu}$, higher confining pressure is needed for the brittle-ductile transition.

Obviously, to ensure the rock exhibit brittle behavior under triaxial pressure, there must be $\bar{\sigma}_3^* > 0$. In addition, according to Eq. (4.4), this requires:

$$\bar{\sigma}_c \geq 2\beta(\sqrt{1+\mu^2}+\mu) \tag{4.6}$$

Equation (4.6) indicates that the ratio $\bar{\sigma}_c$ of rock uniaxial compressive strength σ_c to uniaxial tensile strength σ_t is an important parameter to evaluate rock brittleness. The greater this ratio is, the more brittle the rock is, and vice versa.

References

Atkinson BK (1992) Rock fracture mechanics (trans: Yin X, Xiu J et al). Seismological Press

Brace WF, Paulding BW, Scholz C (1966) Dilatancy in the fracture of crystalline rocks. J Geophys Res 71:3939–3953

Carniglia SC (1972) Reexamination of experimental strength-vs-grain size data for ceramics. J Am Ceram Soc 55(5):243–249

Evans B, Frederich JT, Wong TF (1990) The brittle-ductile transition in rocks: recent experimental and theoretical progress. In: Duba AG, Durham WB, Handin JW, Wang HF (eds) The brittle-ductile transition in rocks. The Heard Volume: American Geophysical Union, Washington DC, Geophys. Monograph 56:1–20

Freiman SW (1984) Effects of chemical environments on slow crack growth in glasses and ceramics. J Geophys Res 89:4027–4076

Gowd TN, Rummel F (1980) Effect of confining pressure on the fracture behaviour of a porous. Int J Rock Mech Sci Geomech Abstr 12:225–229

Griffith AA (1921) The phenomena of rupture and flow in solids. Philos Trans R Soc Ser A Lond 221:163–198

Handin J (1953) An application of high pressure in geophysica: experimental rock deformation. Trans Am Soc Mech Eng 75:315–324

Handin J (1966) Strength and ductility. In: Handbook of physical constants, revised edn. In: Clark SP (ed) Geol Soc Am Memoir 97:223–289

Heard HC (1960) Transition from brittle fracture to ductile flow in Solnhofen limestone as a function of temperature, confining pressure, and interstitial fluid pressure. In: Griggs D, Handin J (eds) Rock deformation. Geol Soc Am Mem 79:193–226

Hou ZM (2003) Mechanical behaviour of salt in the excavation disturbed zone around underground facilities. Int J Rock Mech Min Sci 40(5):727–740

Inglis CE (1913) Stress in a plate due to the presence of cracks and sharp corners. Trans Inst Naval Archit 55:219

Irwin GR (1957) Analysis of stresses and strains near the end of a crack traversing a plate. J Appl Mech 24:361–364

Jaeger JC, Cook NGW, Zimmerman RW (2007) Fundamentals of rock mechanics, 4th edn. Blackwell, Oxford

Johnson B, Gangi AF, Handin J (1978) Thermal cracking of rock subject to show, uniform temperature changes. In: Proceedings of the 19th US symposium on rock mechanics, pp 259—267
Kwasniewski M (1989) Laws of brittle failure and of B-D transition in sandstone. In: Maury V, Fourmaintrax D (eds) Rock at great depth. A. A. Balkema, Rotterdam, pp 45–58
Lawn BR (1993) Fracture of brittle solids, 2nd edn. Cambridge University Press, Cambridge
Li H, Yin G, Xu J, Zhang W (1988) Rock fracture mechanics. Chongqing University Press
Meissner R, Kusznir NJ (1987) Crustal viscosity and the reflectivity of the lower crust. Ann Geophys 5B:365–373
Meng ZP, Peng SP, Qu HL (2000) The relationship between rock composition and structure of roof and floor of coal seam and its mechanical properties. Chin J Rock Mech Eng 19(2):136–139
Michalske TA, Freiman SW (1982) A molecular interpretation of stress corrosion in silica. Nature 295:511–512
Mogi K (1965) Deformation and fracture of rocks under confining pressure. Elasticity and plasticity of some rocks. Bull Earthq Res Inst Tokyo Univ 43:349–379
Mogi K (1966) Pressure dependence of rock strength and transition from brittle fracture to ductile flow. Bull Earthq Res Inst Tokyo Univ 44:215–232
Mogi K (2007) Experimental rock mechanics. Taylor & Francis, Balkema, pp 37–48
Müller L (1981) Rock mechanics (trans: Li S). China Coal Industry Publishing House, Beijing
Nicolas A, Poirier JP (1976) Crystalline plasticity and solid state flow in metamorphic rocks. Wiley, New York, London
Nicholson RB (1972) Electron microscopy and structure of materials. In: Thomas G (ed) University of California Press, Berkeley, p 689
Obrimoff JW (1930) The splitting strength of mica. Proc R Soc Lond 127:290. Bažant ZP, Planas J (1998) Fracture and size effect in concrete and other quasibrittle materials. CRC Press, U.S.A.
Paterson MS (1958) Experimental deformation and faulting in Wombeyan marble. Bull Geol Soc Am 69:465–467
Paterson MS (1982) Experimental rock deformation-the brittle field (Chinese transition). Geological Publishing House
Paterson MS, Wong TF (2005) Experimental rock deformation-the brittle field. Springer, Berlin
Ranalli G, Murphy DC (1987) Rheological stratification of the lithosphere. Tectonophysics 132:281–295
Rice RW, Freiman SW (1981) Grain-size dependence of fracture energy in ceramics II: a model for noncubic materials. J Am Ceram Soc 64(6):350–354
Sibson RH (1977) Fault rock sand fault mechanism. J Geol Soc Lond 13:191–213
Sibson RH (1980) Power dissipation and stress levels on faults in the upper crust. J Geophys Res 85:6239–6247
Shimada M (1993) Lithosphere strength inferred from fracture strength of rocks at high confining pressures and temperatures. Tectonophysics 217:55–64
Watts BM, Stowell MJ, Baikie BL, Owen DGE (1976) Superplasticity in Al-Cu-Zr alloys EM DASH 1. Material preparation and properties. Metal Sci 10(6):197–198
Wiederhom SM (1978) Fracture mechanics of ceramics. In: Bradt RC, Hasselman DPH, Lange FF (eds) vol 4. Plenum Press, New York, pp 549–580
Wong TF, David C, Zhu W (1997) The transition from brittle faulting to cataclastic flow in porous sandstones: mechanical deformation. J Geophys Res 102:3009–3025
Xu X (1998) Preliminary study on mechanical properties and damage characteristics of three gorges granite under temperature action. Institute of Rock and Soil Mechanics, Chinese Academy of Sciences, Wuhan
Xu X, Liu Q (2000) A preliminary study on basic mechanical properties for granite at high temperature. Chin J Rock Mech Eng 22(3):332–335
Zuo JP, Liu HH, Li HT (2015) A theoretical derivation of the Hoek-Brown failure criterion for rock materials. J Rock Mech Geotech Eng 7(4):361–366

Chapter 5
The Hoek-Brown Constant m_i

Abstract The constant m_i is a fundamental parameter required for the Hoek-Brown (HB) failure criterion in estimating the strength of rock materials. In order to calculate m_i values triaxial tests need to be carried out, however, triaxial tests are time-consuming and expensive. Therefore, simplified models are proposed by many researchers to estimate m_i values using rock properties that are easily obtained at an early stage of a project when triaxial test data are not available. These models are in general proposed using statistical technique and the reliability of prediction relies on the quantity and quality of the data used. In this Chapter, existing models (Guidelines, R index, UCS based model and tensile strength based model) are reviewed and compared and their prediction performances are assessed using the published triaxial test data of five common rock types. The estimated intact rock strength is then compared with the experimental intact rock strength using existing triaxial test data. Results show that m_i values calculated from the UCS and TS based model can be reliably used in the HB criterion for estimating the intact rock strength when triaxial test data are not available.

5.1 Introduction

The constant m_i is a fundamental parameter required for the Hoek-Brown (HB) failure criterion in estimating the strength of rock materials. In order to calculate m_i values triaxial tests need to be carried out, however, triaxial tests are time-consuming and expensive. Therefore, simplified models are proposed by many researchers to estimate m_i values using rock properties that are easily obtained at an early stage of a project when triaxial test data are not available. These models are in general proposed using statistical technique and the reliability of prediction relies on the quantity and quality of the data used. In this Chapter, existing models (Guidelines, R index, UCS based model and tensile strength based model) are reviewed and compared and their prediction performances are assessed using the published triaxial test data of five common rock types.

J. Zuo and J. Shen, *The Hoek-Brown Failure criterion—From theory to application*,
https://doi.org/10.1007/978-981-15-1769-3_5

The Hoek-Brown (HB) failure criterion for intact rocks is a non-linear criterion which has been used in various aspects of rock engineering. The failure criterion is expressed in terms of major and minor principal stresses (σ_1 and σ_3), and the mathematical equation is expressed in Eq. (5.1).

$$\sigma_1 = \sigma_3 + \sigma_{ci}\left(m_i \frac{\sigma_3}{\sigma_{ci}} + 1\right)^{0.5} \tag{5.1}$$

The equation contains two intact rock properties, namely, the uniaxial compressive strength (UCS) of the intact rock σ_{ci} and the HB constant m_i. Values of m_i depend on many factors, such as mineral compositions, grain sizes and cementation of rocks. In general, m_i is considered as a curve-fitting parameter for getting the HB failure envelope. However, researches by Zuo et al. (2008, 2015) showed that m_i is not a curve fitting parameter, but has physical meanings and can be derived from micro-mechanics principles (Hoek and Martin 2014). The referred m_i model is expressed in Eq. (5.2).

$$m_i = \frac{\mu \sigma_{ci}}{\beta \sigma_t} \tag{5.2}$$

where σ_{ci} and σ_t are the uniaxial compressive strength and tensile strength (TS) of intact rocks, respectively; μ is the coefficient of friction for the pre-existing sliding crack surfaces, and β is an intermediate fracture mechanics parameter that can be obtained from experimental data.

As m_i values have a significant influence on the rock failure envelope, in order to get the best fit of the failure envelope of a rock, Hoek and Brown (1997) suggested that values of m_i should be calculated over a confining stress σ_3 range from 0 to $0.5\sigma_{ci}$ by using regression analysis. Given that m_i values are calculated from statistical analysis, the quantity and quality of testing data will influence the value of m_i. Researches by Shen and Karakus (2014) indicated that the range of the confining stress σ_3 has a significant influence on m_i values. On the other hand, because triaxial tests are expensive, troublesome and require high-quality core samples, regression analysis is not frequently employed at an early stage of a project and thus simplified methods are required for the preliminary estimation (Cai 2010).

In the absence of triaxial tests, there are five methods available to estimate m_i values, namely, Guidelines (Hoek and Brown 1980; Hoek 2007), R index (Hoek and Brown 1980; Mostyn and Douglas 2000; Cai 2010; Sari 2010; Richards and Read 2011; Read and Richards 2014), UCS based model (Shen and Karakus 2014; Vásárhelyi et al. 2016), TS based model (Wang and Shen 2017) and Crack initiation stress based model (Cai 2010).

Guidelines, R index, UCS and TS based models are in general derived by using statistical methods, and the reliability of the estimation of these four methods depends on the quantity and quality of data used in the analysis. As a consequence, large discrepancies in the predicted m_i values using different methods can be encountered, which will reduce the confidence of the estimated rock strength values.

In this Chapter, the four existing methods (Guidelines, R index, UCS and TS based models) for estimating m_i values are reviewed and evaluated by using 112 groups of triaxial test data obtained from 'RocData' for five common rock types.

5.2 Indicators for the Evaluation of Existing Methods

To evaluate the prediction performance of existing methods, an extensive database of triaxial tests for five common rock types is selected from 'RocData' as indicated in Table 5.1.

To assess the reliability of existing methods (Guidelines, R index, UCS and TS based model), the estimated rock strength σ_{1_est} under confining stress σ_{3_est} for a given group of triaxial data (σ_{3_est}, σ_{1_est}) is calculated from the HB criterion using m_i values estimated from different methods. Then discrepancies between the estimated rock strength (σ_{1_est}) and the testing rock strength (σ_{1_test}) are indicated by the coefficient of determination (R^2), the absolute average relative error percentage (AAREP) and the maximum discrepancy (M_d). R^2 and AAREP characterize the discrepancies between σ_{1_est} and σ_{1_test} for a group of data; M_d illustrates the maximum value of AAREP among different groups of data for a specific rock type. More specifically, the smaller AAREP values or M_d values are, the smaller discrepancies exist between σ_{1_est} and σ_{1_test}, the higher the prediction performance of the method has. Definitions of the indicators are shown as follows:

$$R^2 = 1 - \frac{\sum_{i=1}^{N} \left(\sigma_{1\_test}^{i} - \sigma_{1\_est}^{i}\right)^2}{\sum_{i=1}^{N} \left(\sigma_{1\_test}^{i} - \overline{\sigma}_{1\_test}\right)^2} \tag{5.3}$$

$$D_p = \frac{\sigma_{1\_est}^{i} - \sigma_{1\_test}^{i}}{\sigma_{1\_test}^{i}} \times 100\% \tag{5.4}$$

$$\text{AAREP} = \frac{\sum_{i=1}^{N} |D_p|}{N} \tag{5.5}$$

$$M\text{d} = \text{Max}(\text{AAREP}) \tag{5.6}$$

Table 5.1 Triaxial tests of five rock types in the database

Rock type	Data groups	Data sets
Coal	32	176
Granite	12	103
Limestone	21	119
Marble	15	121
Sandstone	32	277

where N equals the number of testing data used; $\bar{\sigma}^{i}_{1\_test}$ and $\bar{\sigma}^{i}_{1\_test}$ refer to the estimated rock strength and the experimental rock strength respectively; $\bar{\sigma}_{\_test}$ is the mean value of the experimental σ_{1_test} values. It is known that the HB failure envelope can be divided into the tensile zone where $\sigma_3 < 0$ and the compression zone where $\sigma_3 > 0$. In order to generate the best fit of HB failure envelope, Cai (2010) suggested to use two different m_i values to fit the test data; one is $(m_i)_t$ for tension zone and the other is $(m_i)_c$ for the compression zone. Carter et al. (1991) also noted that the HB failure envelope does not fit the test data quite well in tension zone. In this study, the prediction performance of existing models and proposed model was assessed within triaxial test data in the compression zone and therefore it should be noted that TS and UCS points are not included in the calculation of these indicators.

5.3 Existing Methods for Estimating m_i Values

5.3.1 *Regression Method*

In order to obtain the best rock strength prediction, Hoek and Brown (1997) suggested that the values of m_i should be calculated over a confining stress σ_3 range from 0 to 0.5 σ_{ci} by using regression methods (Eqs. 5.7 and 5.8), and at least five sets of triaxial tests should be included in the regression analysis.

$$\sigma_{ci\_fitted} = \sqrt{\frac{\sum y}{n} - \left[\frac{\sum xy - \left(\sum x \sum y / n\right)}{\sum x^2 - \left(\left(\sum x\right)^2 / n\right)}\right] \frac{\sum x}{n}} \tag{5.7}$$

$$m_i = \frac{1}{\sigma_{ci\_fitted}} \left[\frac{\sum xy - \left(\sum x \sum y / n\right)}{\sum x^2 - \left(\left(\sum x\right)^2 / n\right)}\right] \tag{5.8}$$

where, $x = \sigma_3$, $y = (\sigma_1 - \sigma_3)^2$, n is the number of triaxial data sets. It should be noted that the value of UCS in Eq. 5.7 is calculated from triaxial data, which is generally different from UCS estimated from the uniaxial compression test. In general, if regression analysis includes data points with high confining pressure to get the UCS, the fitted UCS will be significantly higher than that obtained from the uniaxial compression test.

Read and Richards (2011) suggested that the most accurate method of assessing m_i values is regression analysis (including triaxial, uniaxial compressive and tensile tests) within the confining stress range from σ_t to $0.5\sigma_{ci}$. Given that m_i values calculated from statistical analysis, the reliability of the calculated m_i values depends on the quantity and quality of testing data used in the regression method.

Research by Singh et al. (2011) indicated that the range of confining stress σ_3 can have a significant influence on the calculation of m_i.

Table 5.2 shows a comparison study on the calculation of m_i values from different confining stress σ_3 for limestone conducted by Schwartz (1964). For example, if the first three data sets are selected m_i would be equal to 5.16. However, if all eleven data points are used m_i value will then be 1.21. The value of T in Table 5.2 represents the ratio of the maximum to the minimum of m_i value ($T \geq 1$), for example, $T = m_{i\_max}/m_{i\_min} = 5.16/1.21 = 4.26$ in this case. T equaling to 1 means that there is no variation in the parameter m_i, although different sets of data are used.

After obtaining the values of m_i, the strength of intact rock under a given value of confining stress σ_3 can be calculated from Eq. (5.1). The estimated rock strength is then compared with that from the experimental data.

The HB failure envelopes using various m_i values are plotted in Fig. 5.1, and their prediction performances (indicated by R^2 and AAREP) are shown in Table 5.2. For a given value of m_i, all the triaxial tests (exclude the UCS point) were used to calculate the values of AAREP and R^2.

Results of Table 5.2 shows that the rock strength prediction of the HB criterion will significant influenced by the m_i value. To extend the analysis to other triaxial tests for various rock types, the sensitivity of m_i was tested using the database indicated in Table 5.1. The histogram in Fig. 5.2 shows T values distribution for a complete comparison.

Based on the assessment of T, the results illustrate that the parameter m_i has high sensitivity to variations in confining stress σ_3. 33% of the data sets have T values greater than 2.5. This statistical analysis indicates that discrepancies in the predicted values of m_i using different sets of test data can result in reducing confidence in the predicted rock strength values. From this perspective, the best way to achieve the best fit can be the use of optimal m_i values that minimize AAREP or R^2 using

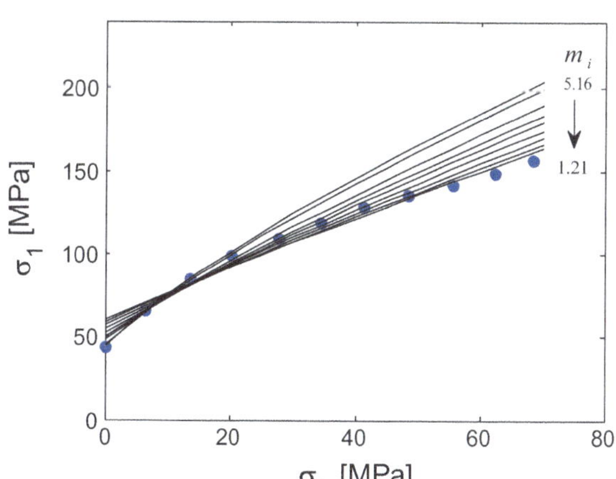

Fig. 5.1 The Hoek-Brown failure envelopes using different m_i values

Table 5.2 Estimated mi values by regression analysis using triaxial test data at different confining stresses

Data sets (N)	σ_3 (MPa)	σ_1 (MPa)	m_i	R^2	AAREP (%)
1	0	44	–	–	–
2	6.5	66	–	–	–
3	13.7	85	5.16	0.17	14.24
4	20.3	99	4.63	0.36	12.28
5	27.9	109	3.53	0.67	8.79
6	34.4	119	2.97	0.79	6.87
7	41.2	128.2	2.54	0.86	5.57
8	48.4	135.1	2.07	0.92	4.67
9	55.4	141.9	1.68	0.95	4.28
10	62.3	149.1	1.40	0.97	4.00
11	68.4	156.5	1.21	0.97	3.99
Optimal m_i			1.31	0.97	3.84
T			4.26		

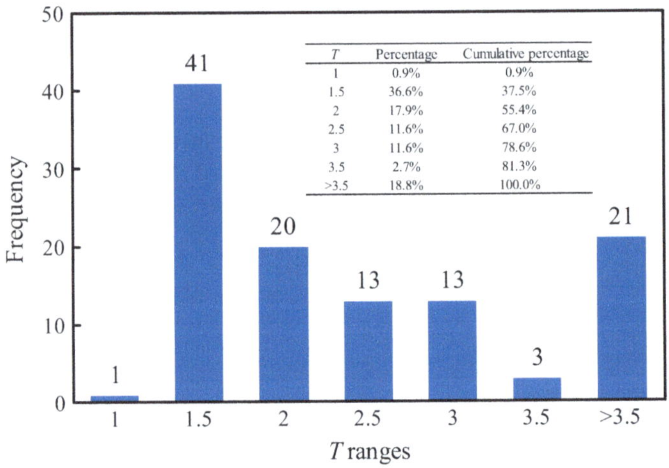

Fig. 5.2 Comparison of sensitivities to the confining stress range employed for m_i fitting, as indicated by the T parameter

optimization solution routines, such as 'Excel Solver'. The optimal m_i value for the triaxial data in the case in Table 5.2 is equal to 1.31, which has the lowest value of AAREP = 3.84%.

5.3.2 Guidelines

Values of m_i depend on many factors, such as grain sizes and mineral compositions. In other words, m_i reflects rock types and based on which, Hoek and Brown (1980) proposed Guidelines for pre-determining m_i values for different rock types. They suggested that values of m_i increase with rock types in the following general ways: (1) $m_i = 7$ for carbonate rocks with well-developed crystal cleavage, such as dolomite, limestone and marble; (2) $m_i = 10$ for lithified argillaceous rocks, such as mudstone, siltstone, shale and slate; (3) $m_i = 15$ for arenaceous rocks with strong crystals and poorly developed crystal cleavage, such as sandstone and quartzite; (4) $m_i = 17$ for fine-grained polyminerallic igneous crystalline rocks, such as andesite, dolerite, diabase and rhyolite; (5) $m_i = 25$ for coarse-grained polyminerallic igneous and metamorphic rocks, such as amphibolite, gabbro, gneiss, granite, norite and quartz-diorite. The latest version of Guidelines was proposed by Hoek (2007) which associates with a more detailed lithological classification. Table 5.3 shows the suggested m_i ranges for different rock types.

The correlations between m_i values based on Guidelines and m_i values calculated from the regression analysis were extensively tested by Mostyn and Douglas (2000) and Read and Richards (2014) for a variety of rock types. Their results showed that the correlations between Guidelines and regression based m_i are not quite strong; generally the range of calculated m_i values using regression analysis has a much greater spread than that in Guidelines and about 50% of testing data fall outside the Guidelines range for each rock type. Comparisons of m_i values obtained from these two methods for five common rock types are shown in Fig. 5.3. The value of m_i of guidelines for coal was obtained from Hoek and Brown (1997), and the m_i values for the other four rock types were obtained from Hoek (2007). Such a large variation range presents a major challenge for engineers to choose a reasonable m_i value for a particular rock type.

5.3.3 R Index

R index, the ratio of the uniaxial compressive strength σ_{ci} to the tensile strength σ_t, was also suggested by many researchers (Hoek and Brown 1980; Mostyn and Douglas 2000; Cai 2010; Sari 2010; Richards and Read 2011; Read and Richards 2014) as an alternative way to assess m_i values in the absence of triaxial test data. It is known that direct tensile tests are not routinely carried out as a standard procedure in many rock testing laboratories because of the difficulty in the specimen preparation. Indirect methods, on the other hand, such as Brazilian tests are widely used to estimate the tensile strength. The correlations between m_i values and R index which was calculated using direct tensile and Brazilian tests are compared as shown in Fig. 5.4. The solid diagonal line in Fig. 5.4 represents $R = m_i$. The upper and lower dash lines represent the six over-estimate and under-estimate of m_i values

Table 5.3 Guidelines-based values of the constant mi for intact rocks (Hoek 2007)

Rock type	Class	Group	Texture			
			Coarse	Medium	Fine	Very fine
Sedimentary	Clastic		Conglomerates 21 ± 3 Breccias 19 ± 5	Sandstones 17 ± 4	Siltstones 7 ± 2 Greywackes 18 ± 3	Claystones 4 ± 2 Shales 6 ± 2 Marls 7 ± 2
	Non-Clastic	Evaporites		Gypsum 8 ± 2	Anhydrite 12 ± 2	
		Carbonates	Crystalline Limestone 12 ± 3	Sparitic Limestones 10 ± 2	Micritic Limestones 9 ± 2	Dolomites 9 ± 3
		Organic	Coal[a] 8–21			Chalk 7 ± 2
Metamorphic	Non foliated		Marble 9 ± 3	Hornfels 19 ± 4 Metasandstone 19 ± 3	Quartzites 20 ± 3	
	Slightly foliated		Migmatite 29 ± 3	Amphibolites 26 ± 6		
	Foliated		Gneiss 28 ± 5	Schists 12 ± 3	Phyllites 7 ± 3	Slates 7 ± 4
Igneous	Plutonic	Light	Granite 32 ± 3	Diorite 25 ± 5		
			Granodiorite 29 ± 3			
		Dark	Gabbro 27 ± 3	Dolerite 16 ± 5		
			Norite 20 ± 5			
	Hypabyssal		Porphyries 20 ± 5		Diabase 15 ± 5	Peridotite 25 ± 5
	Volcanic	Lava		Rhyolite 25 ± 5 Andesite 25 ± 5	Dacite 25 ± 3 Basalt 25 ± 5	Obsidian 19 ± 3
		Pyroclastic	Agglomerate 19 ± 3	Breccia 19 ± 5	Tuff 13 ± 5	

[a]The guidelines-based m_i value for coal was obtained from Hoek and Brown (1997)

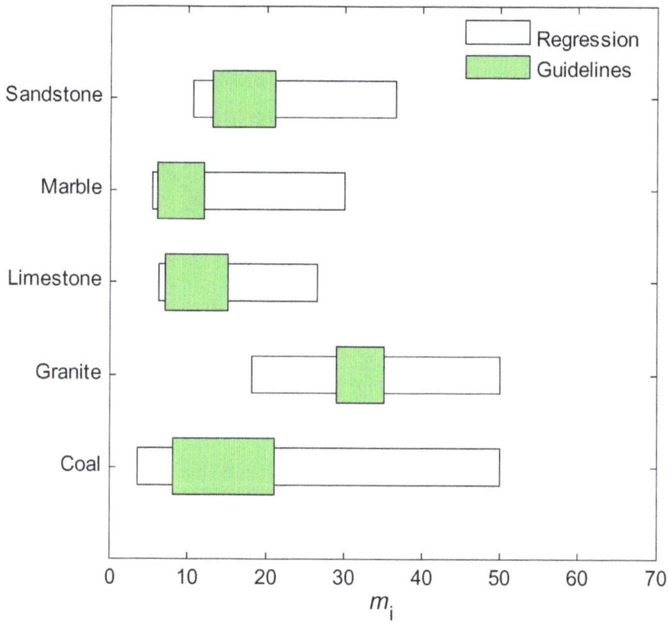

Fig. 5.3 m_i values based on regression analysis and guidelines for five rock types

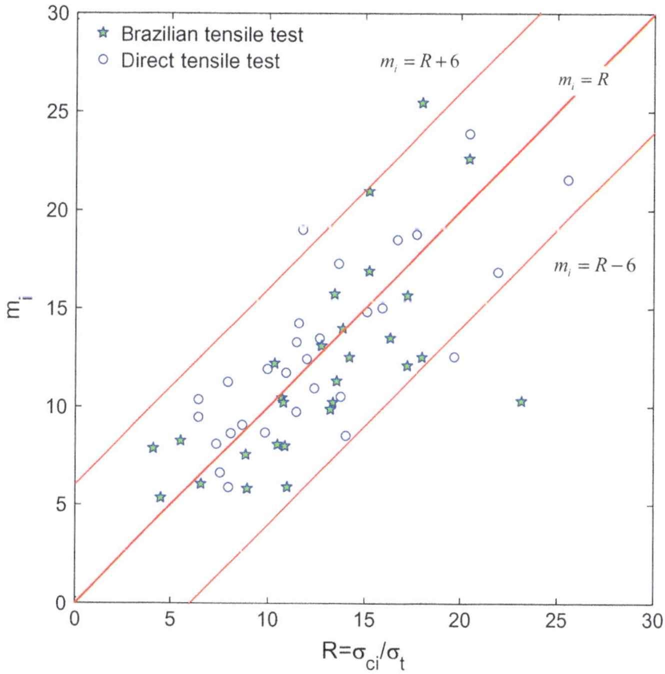

Fig. 5.4 Correlation between R and m_i, after Richards and Read (2011)

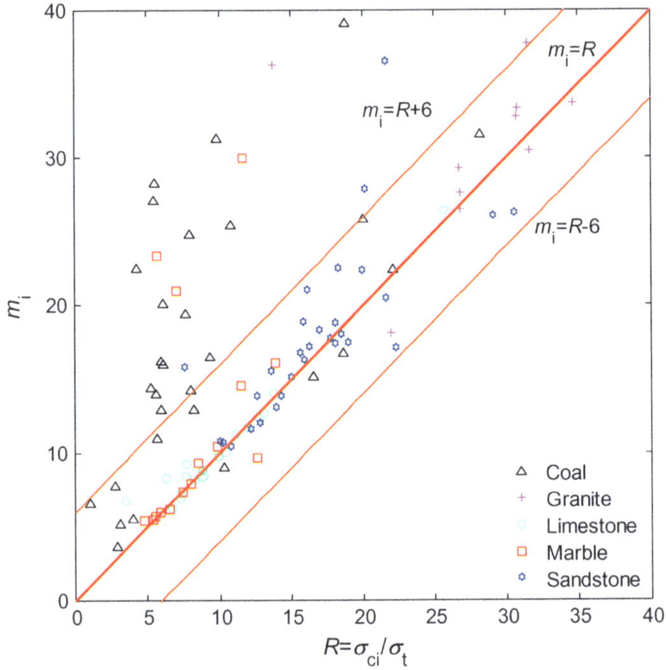

Fig. 5.5 Correlation between R and m_i for 5 rock types (Shen and Karakus 2014)

respectively. The results illustrate that only 4 out of 57 sets of data fall out of the lines between m_i = R±6, which suggests that the absolute error of m_i values is ±6 and R index has a quite high-level of confidence. In addition, based on the analysis of 44 groups laboratory data, Read and Richards (2014) found that direct tensile strength (DTS) values may be taken as 0.9 times of the Brazilian tensile strength (BTS) values and suggested that R can be estimated as UCS/(0.9BTS) using laboratory test values from uniaxial compression and Brazilian tensile tests.

The correlations between R index and m_i values were also compared for different rock types using the database presented in Table 5.1 as shown in Fig. 5.5. Generally, except for coal, most of the data fall between the lines m_i = R±6.

5.3.4 UCS Based Model

Based on the analysis of theoretical relationship between UCS and m_i which was calculated from the regression analysis (see Eqs. 5.7 and 5.8) proposed by Hoek and Brown (1997), the Shen and Karakus (2014) proposed a simplified method to estimate m_i values directly from the UCS for a specific rock type when triaxial test data are not available. Equation (5.8) shows that the value of m_i will decrease with the increase of UCS. Although the relation between fitted UCS and m_i in Eq. (5.8) is based on triaxial data, it gives them the inspiration to investigate the possible relation between m_i and UCS which is obtained from the uniaxial compression test.

Additionally, according to the field estimates of UCS of intact rock presented by Hoek and Brown (1997) there exist general correlations between the UCS and rock types, for example chalk (5–25 MPa), coal (25–50 MPa), sandstone (50–250 MPa) and granite (>250 MPa). The field estimates of UCS and the guidelines method to estimate m_i values indicates that the values of UCS and m_i have relations with rock type. Therefore, the authors were interested in investigating the potential relationship between m_i and UCS for a given specific rock type.

Considering that the UCS of intact rocks is one of the most important rock properties in rock engineering applications and can be relatively straightforwardly estimated in a cost-effective way, Shen and Karakus (2014) proposed a new method to estimate m_i values directly from the UCS of intact rocks when triaxial test data are not available. The correlation is presented in Eq. (5.9).

$$m_i = a\sigma_{ci}^{b+1} \tag{5.9}$$

where σ_{ci} represents the uniaxial compressive strength of intact rocks; a and b are constants depending on rock types. The correlations between σ_{ci} and m_i values are presented in Fig. 5.6.

Vásárhelyi et al. (2016) used the UCS based model to evaluate the m_i values of Hungarian granitic rocks of Bátaapáti site, Hungary. The database contains 44 samples which represent three lithologies: 15 samples of monzonitic, 14 samples of monzogranitic and 15 samples of hybrid rocks (contaminated monzonite and monzogranite). Their results showed that values of m_i decrease with the increase of UCS values of rocks are in a good agreement with the findings of Shen and Karakus (2014). Values of a and b calculated by Shen and Karakus (2014) and Vásárhelyi et al. (2016) for different rock types are shown in Table 5.4.

5.3.5 TS Based Model

The tensile strength (TS) of intact rocks is one of the most important rock properties in rock engineering applications. Many researchers found the correlations between TS and other rock properties, such as point load index (Kılıç and Teymen 2008) and crack initiation threshold (Diederichs and Martin 2010). A recent study performed by Karaman et al. (2015) showed that there is a strong linear correlation between the MC shear strength parameter cohesion and the tensile strength of intact rocks.

Based on the fact that both the MC parameter cohesion c and the HB constant m_i are usually estimated from triaxial test data using appropriate curve-fitting strategies, therefore, we were interested in investigating the potential correlations between m_i values and TS for different rock types.

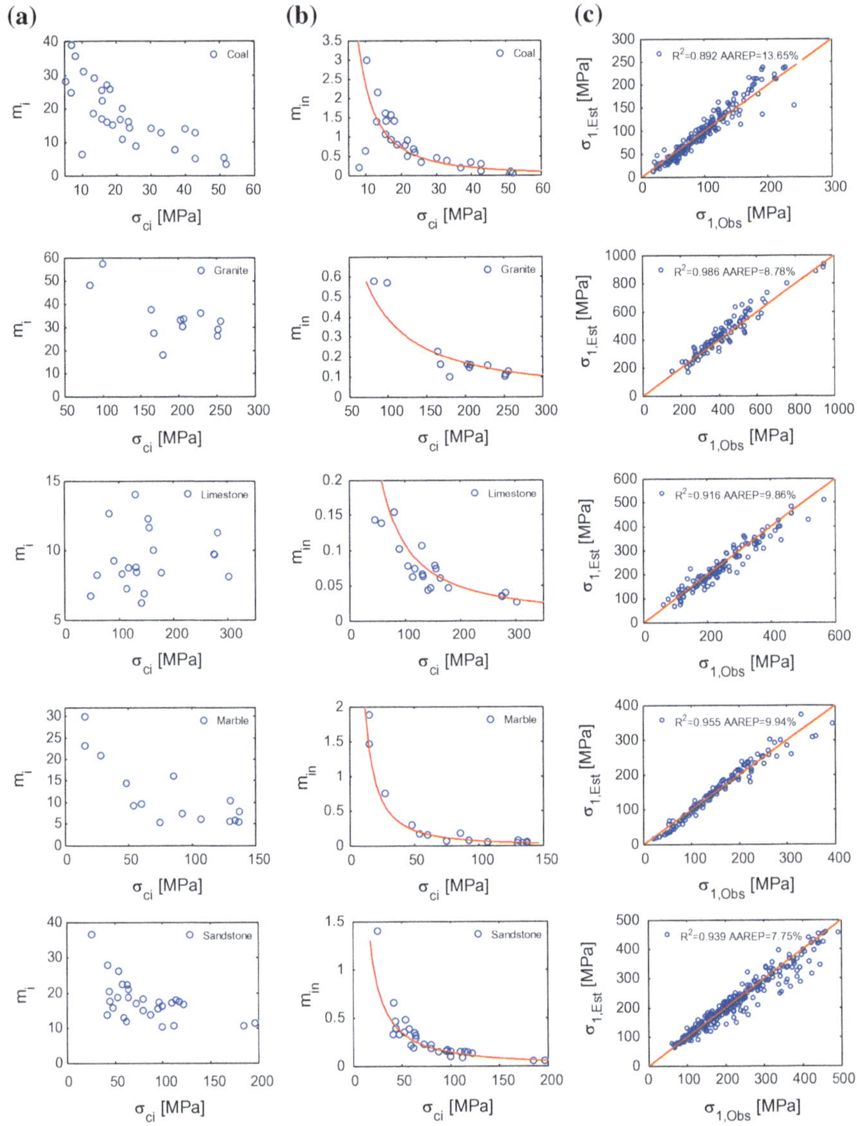

Fig. 5.6 **a** Correlations between m_i and σ_{ci}. **b** Correlations between m_{in} and σ_{ci}. **c** Rock strength prediction performances

Such correlations between m_i values and TS for five common rock types are shown in Fig. 5.7. Values of m_i were calculated from regression analysis over a confining stress σ_3 range from 0 to $0.5\sigma_{ci}$ as suggested by Hoek and Brown (1997).

Figure 5.7 shows that there is a trend of decreasing m_i values with the increasing tensile strength. However, such correlations for limestone are not observed. This

Table 5.4 Best fit a and b constants for UCS based model

Rock type	a	b
Coal	120	−1.70
Granite	100	−1.20
Limestone	22	−1.15
Marble	100	−1.55
Sandstone	50	−1.26
Monzonite	120	−1.40
Monzogranite	87	−1.14
Hybrid rocks	387	−1.67

may be because limestone has different compositions and cementations, which leads to the data widely scattered. Equation (5.10) can be used to calculate m_i values from TS for five different rock types by fitting the curve. TS based model is used to refer to this model in the following sections.

$$m_i = A\sigma_t^B \tag{5.10}$$

where A and B are constants depending on rock types. In general, the constants A and B are a curve-fitting parameter. However, according to the theory of Zuo et al. (2015) the m_i value can be estimated from the coefficient of friction for the pre-existing sliding crack surfaces, and fracture mechanics parameter. From this point of view, the constants A and B for the TS based model seems to have its physical meaning as well, which have possible relations with the micro-mechanics principles. However, investigating the possible relations between these constants and other rock properties and explaining the possible physical meanings need to carry out further research to get reliable and enough data. The best fit for five rock types of A and B are shown in Table 5.5, which is used to calculate m_i values for the evaluation of the rock strength prediction for each type of rocks in the next section (as indicated by R^2, AAREP and M_d values).

It should be noted that the database obtained from 'RocData' did not provide information about whether the tensile strength is the direct tensile strength (DTS) or the Brazilian tensile strength (BTS) and therefore the type of the tensile strength data used in the study was not emphasized. However, based on the analysis of 44 groups of data, Read and Richards (2014) found that the direct tensile strength (DTS) values may be taken as 0.9 times of the Brazilian tensile strength (BTS) values, which means the type of the tensile strength may affect the accuracy of proposed TS based model.

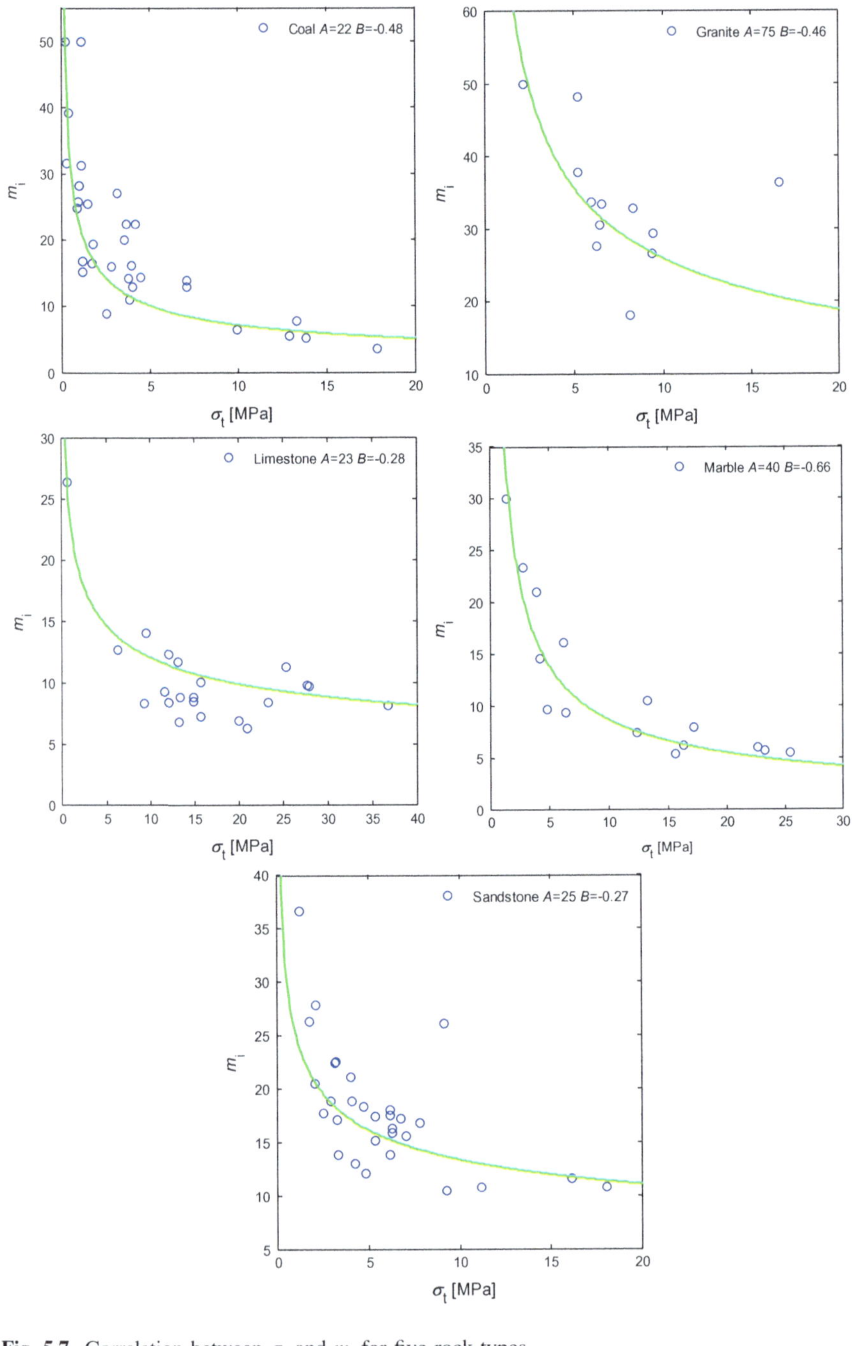

Fig. 5.7 Correlation between σ_t and m_i for five rock types

Table 5.5 Best fit A and B for TS based model

Rock type	A	B	AAREP (%)	R2	M_d (%)
Coal	22	−0.48	11.65	0.94	27.97
Granite	75	−0.46	7.17	0.98	25.17
Limestone	23	−0.28	8.93	0.93	26.76
Marble	40	−0.66	8.18	0.98	22.12
Sandstone	25	−0.27	7.06	0.96	20.23

5.3.6 Methods Based on Other Rock Properties

Using the cracking initiation stress σ_{ci} and the peak strength σ_f, Cai (2010) proposed a method that depends on the confining pressure to estimate m_i values. The crack initiation stress σ_{ci} can be obtained by the acoustic emission monitoring and the volumetric strain measurement, as shown in Fig. 5.8. The results showed that $m_i \approx 12\sigma_f/\sigma_{ci}$ can be applied to strong, brittle rocks, applicable to the high confinement zone. $m_i \approx 8\sigma_f/\sigma_{ci}$ can be used in the low confinement stress to tension zone.

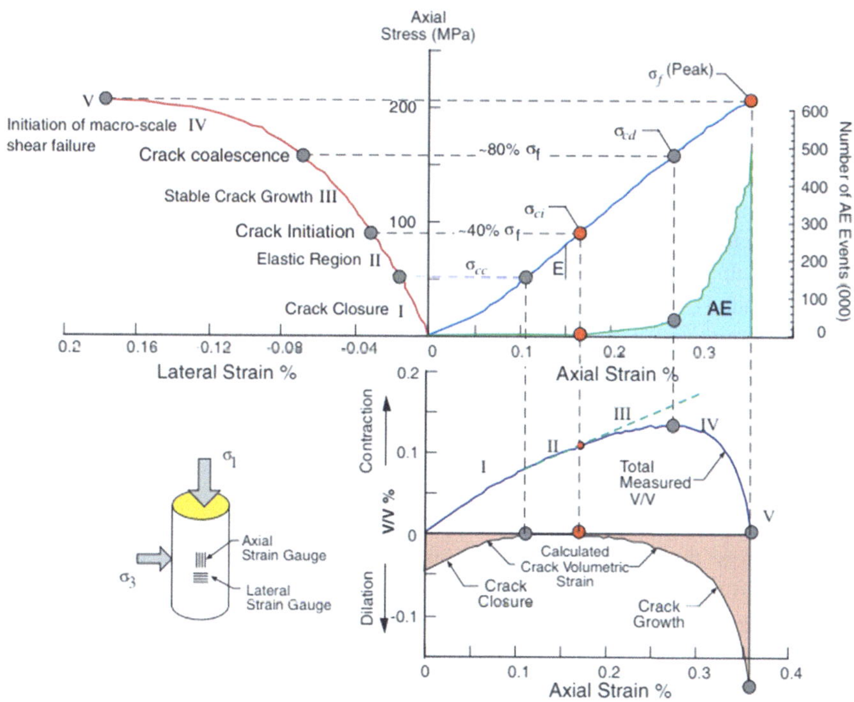

Fig. 5.8 Stress–strain diagram of a granite showing the stages of crack development (Cai 2010)

Additionally, a three-dimensional crack model proposed by Zuo et al. (2008, 2015) is able to theoretically derive m_i from a failure characteristic factor $(\mu\sigma_{ci})/(\beta\sigma_t)$. The contribution of this model is that it links the empirical HB criterion with micro-mechanics principles. They suggested that when macro-failure occurs, a failure characteristic factor, which is a product of the micro-failure orientation angle and the changing rate of the angle with respect to the major principal stress, must exceed a critical value. By using the failure characteristic factor to derive m_i values from micro-mechanical considerations, the model is able to relate the microscopic damaged crack characteristics to macro-failures successfully.

5.4 Comparison of Prediction Performance of Various Models

The prediction performance of various models for five specific rock types is shown in Fig. 5.8 and Table 5.6. In Fig. 5.9, the horizontal axis represents AAREP values calculated from a group of data for a specific rock type. The vertical axis represents the cumulative distribution function (CDF) of AAREP values. In other words, CDF is the probability that AAREP takes a value less or equal to a certain value. With this definition, the curve with a 'higher' position in the plot indicates that the corresponding model provides a better strength prediction.

Results presented in Table 5.6 show that for coal, m_i values estimated by Guidelines and R index generate large discrepancies of rock strength σ_1; AAREP values are higher than 20% and M_d values are up to 53.19 and 48.32%, respectively. Although the UCS based model's AAREP value is smaller than 15%, its M_d value is quite large which is up to 70%. In the absence of triaxial data, it is suggested to adopt TS based model to calculate m_i values, which generates quite acceptable results with the AAREP and M_d values are equal to 11.64 and 27.97%, respectively, and the results of which is close to the results from the regression analysis.

For granite, UCS based model is recommended as it gives the lowest M_d value (16.81%) compared with Guidelines (21.43%), R index (35.81%) and TS based model (25.17%). The value of AAREP (8.79%) of UCS based model is also acceptable. When no testing data are available, Guidelines (AAREP = 8.22% and M_d = 21.43%) is also acceptable at an early stage of practical engineering.

For limestone, the prediction performance of R index and TS based model is quite close; R index has AAREP = 11.45% and M_d = 25.79%, and TS based model has AAREP = 8.93% and M_d = 26.76%. Considering that TS based model doesn't need to know the values of UCS of intact rocks compared with R index, it is suggested to take TS based model as a priority.

For marble, M_d values of Guidelines and R index are larger than 40% while TS or UCS based model gives the relatively good prediction performance. Therefore, lower cost procedures such as point load, uniaxial compressive and Brazilian tests are recommended.

Table 5.6 Prediction performance of five methods for five specific rock types

Methods	Coal			Granite			Limestone			Marble			Sandstone		
	R^{2a}	AAREP[a] (%)	M_d (%)	R^{2a}	AAREP[a] (%)	M_d (%)	R^{2a}	AAREP[a] (%)	M_d (%)	R^{2a}	AAREP[a] (%)	M_d (%)	R^{2a}	AAREP[a] (%)	M_d (%)
Regression	0.89	9.8	23.1	0.98	4.8	9.5	0.98	4.2	7.4	0.98	5.8	13.4	0.99	4.0	12.2
Guidelines	0.73	20.1	53.1	0.99	8.2	21.4	0.93	10.3	27.8	0.93	15.7	42.0	0.93	8.5	26.7
R index	0.82	21.4	48.3	0.96	11.0	35.8	0.90	11.4	25.7	0.96	13.5	43.1	0.95	6.9	26.2
UCS based	0.89	13.6	7033	0.99	8.7	16.8	0.92	10.2	31.1	0.96	9.9	25.1	0.94	7.7	25.6
TS based	0.94	11.6	27.9	0.98	7.1	25.1	0.93	8.9	26.7	0.98	8.1	22.1	0.96	7.0	20.2

[a]R^2 and AAREP used here are calculated from all groups of data for a specific rock type

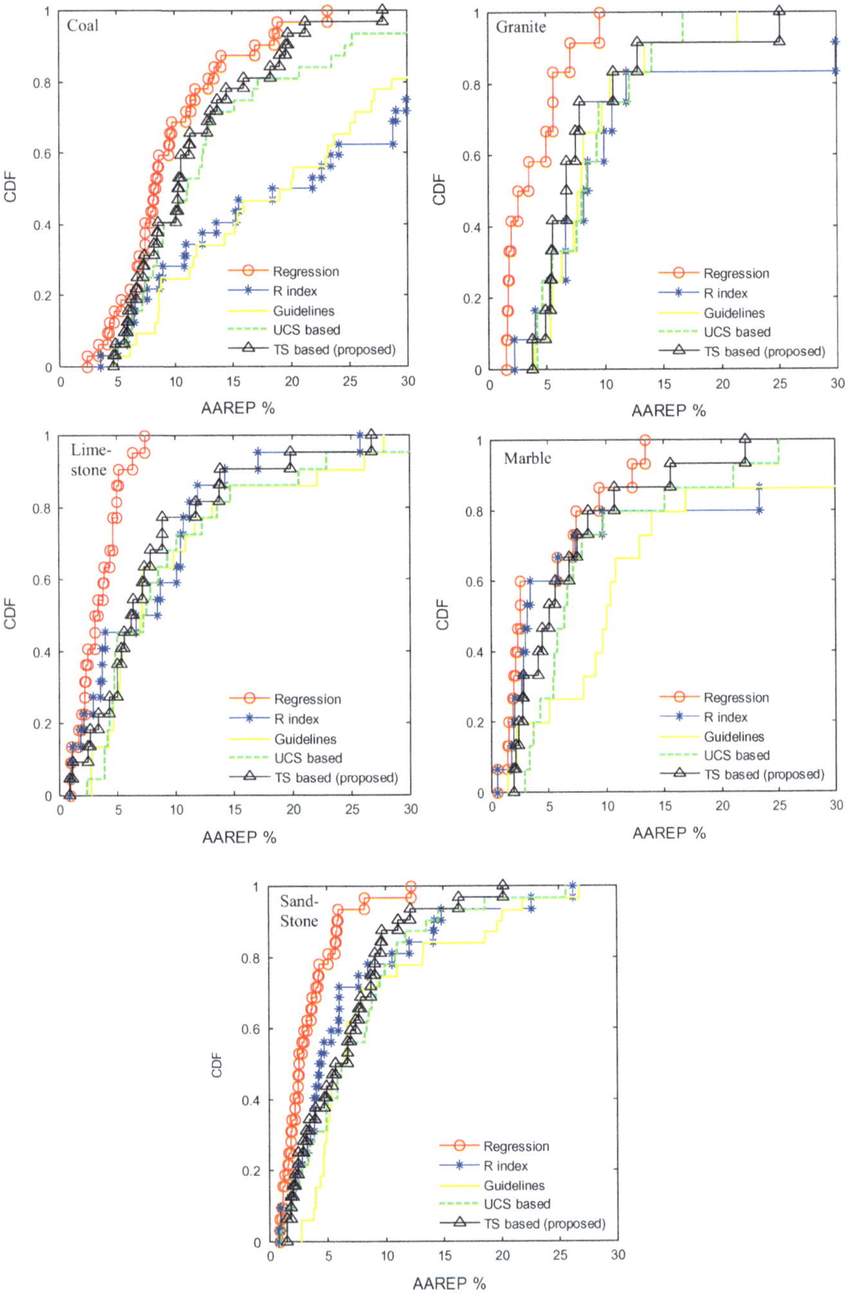

Fig. 5.9 Cumulative distribution function (CDF) of prediction errors (AAREP) of five methods for five specific rock types

For sandstone, values of R^2 and AAREP calculated from, R index, UCS and TS based models are quite close. However, TS based model gives a relatively low value of M_d (20.23%) compared with R index (26.23%) and UCS based model (25.62%). Therefore, TS based model is suggested as a priority to estimate m_i values. In the situation without any tests, Guidelines can also generate reasonable results with AAREP = 8.54% and M_d = 26.77%.

Based on the comprehensive statistical analysis, recommended methods for different rock types are summarized in Table 5.7. It should be noted that for a specific rock type, Guidelines gives a range of m_i values. Therefore, the prediction performance of Guidelines actually depends on the choice of m_i values based on rock physical properties, such as mineral compositions, grain sizes and cementation of rocks. Because of the lack of detailed lithological descriptions of rocks samples in the database, the mean value of m_i based on Guidelines was used in the study.

A comprehensive study of all five rock types is conducted to compare the overall predictive capability of existing and proposed methods. The results are shown in Fig. 5.10 indicate that there is no doubt that regression analysis yields the best prediction performance, while in the absence of triaxial tests, overall, the proposed TS based model gives the best prediction performance compared with other three methods.

Table 5.7 Recommended methods for five specific rock types

Models	Coal	Granite	Limestone	Marble	Sandstone
Guidelines	X	XX	XX	X	XX
R index	X	X	XXX	X	XXX
UCS based	XX	XXXX	XX	XXX	XXX
TS based (proposed)	XXXX	XXX	XXXX	XXXX	XXXX

[a]Methods indicated as 'XXXX' represent prior suggested methods. Methods indicated as 'X' mean the least suggested methods

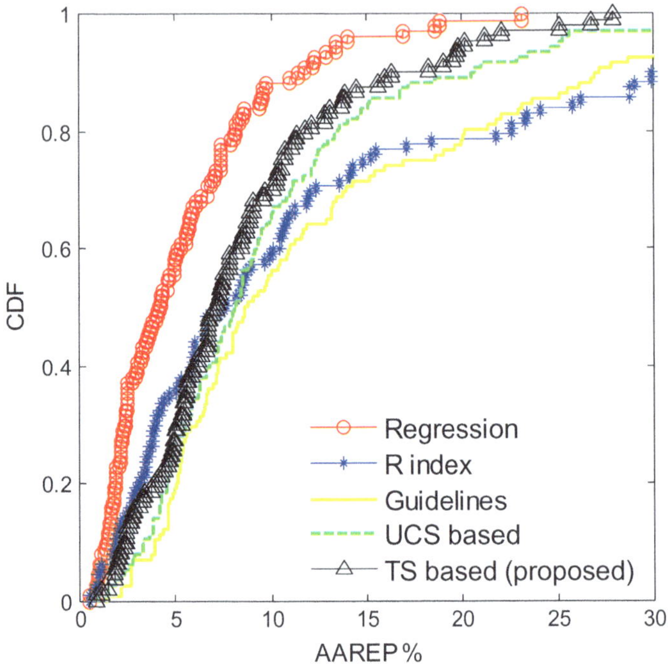

Fig. 5.10 Comparison of the rock strength prediction of the proposed method with existing methods

References

Cai M (2010) Practical estimates of tensile strength and Hoek-Brown strength parameter mi of brittle rocks. Rock Mech Rock Eng 43(2):167–184

Carter B, Duncan E, Lajtai E (1991) Fitting strength criteria to intact rock. Geotech Geol Eng 9:73–81

Diederichs M, Martin C (2010) Measurement of spalling parameters from laboratory testing. In: Proceedings of European rock mechanics symposium. Taylor and Francis, London

Hoek E (2007) Rock mass properties. Chapter 11, Practical rock engineering. www.rocscience.com/learning/hoek/corner/11_Rock_mass_properties.pdf. Accessed 18 Jan 2017

Hoek E, Brown ET (1980) Empirical strength criterion for rock masses. J Geotech Geoenvironmental Eng (ASCE) 106:15715

Hoek E, Brown ET (1997) Practical estimates of rock mass strength. Int J Rock Mech Min Sci 34 (8):1165–1186

Hoek E, Martin CD (2014) Fracture initiation and propagation in intact rock-a review. J Rock Mech Geotech Eng 6(4):287–300

Karaman K, Cihangir F, Ercikdi B, Kesimal A, Demirel S (2015) Utilization of the Brazilian test for estimating the uniaxial compressive strength and shear strength parameters. J South Afr Inst Min Metall 115(3):185–192

Kılıç A, Teymen A (2008) Determination of mechanical properties of rocks using simple methods. Bull Eng Geol Env 67(2):237–244

Mostyn G, Douglas K (2000) Strength of intact rock and rock masses. In: Haberfield C et al (eds) Proceedings of international conference on geotechnical and geological engineering, vol 1. Technomic Publishing, Lancaster, PA, pp 1389–1421

Read S, Richards L (2014) Correlation of direct and indirect tensile tests for use in the Hoek-Brown constant mi. In: Rock engineering and rock mechanics: structures in and on rock masses. Taylor and Francis, London

Richards L, Read S (2011) A comparative study of mi, the Hoek-Brown constant for intact rock material. In: Proceedings 45th US rock mechanics/geomechanics symposium. American Rock Mechanics Association, San Francisco, CA

Sari M (2010) A simple approximation to estimate the Hoek–Brown parameter "mi" for intact rocks. In: Rock mechanics in civil and environmental engineering. Taylor and Francis, London

Schwartz A (1964) Failure of rock in the triaxial shear test. In: Proceedings of the 6th US rock mechanics symposium. Rolla, Mo, pp 109–135

Shen J, Karakus M (2014) Simplified method for estimating the Hoek-Brown constant for intact rocks. J Geotech Geoenvironmental Eng (ASCE) 140(6):04014025

Singh M, Raj A, Singh B (2011) Modified Mohr-Coulomb criterion for non-linear triaxial and polyaxial strength of intact rocks. Int J Rock Mech Min Sci 48:546–555

Vásárhelyi B, Kovács L, Török Á (2016) Analysing the modified Hoek-Brown failure criteria using Hungarian granitic rocks. Geomech Geophys Geo-Energy Geo-Resour 2(2):131–136

Wang W, Shen J (2017) Comparison of existing methods and a new tensile strength based model in estimating the Hoek-Brown constant m_i, for intact rocks. Eng Geol 224:87–96

Zuo J, Li H, Xie H, Ju Y, Peng S (2008) A nonlinear strength criterion for rock-like materials based on fracture mechanics. Int J Rock Mech Min Sci 45(4), 594–599

Zuo J, Liu H, Li H (2015) A theoretical derivation of the Hoek-Brown failure criterion for rock materials. J Rock Mech Geotech Eng 7(4):361–366

Chapter 6
The Geological Strength Index

Abstract A rock mass failure criterion would have no practical value unless it could be linked to geological description that could be made easily by engineering geologists. The geological strength index (GSI) is a rock mass classification system that has been developed in rock engineering to meet the need of estimating reliable rock mass properties for the design of engineering projects. The heart of the GSI system is a careful engineering geology description of the rock mass encountered in engineering projects. The value of GSI is based upon the assessment of the two fundamental factors, namely, structure and discontinuities condition in the rock mass which can be estimated from visual examination of the rock mass exposed in outcrops. This Chapter attempts to introduce various descriptive and quantitative GSI charts for describing rock structure and the surface conditions of the discontinuities.

6.1 Introduction

The geological strength index (GSI) is a rock mass classification system that has been developed in rock engineering to meet the need of estimating reliable rock mass properties for the design of engineering projects, such as tunnels, slopes and foundations. The core of the GSI classification system is a careful engineering geology description of the rock mass which involves two factors, namely, the rock structures and the surface condition of discontinuities. The value of GSI can be estimated from visual examination of the rock mass exposed in outcrops and borehole cores. Once the value of GSI is estimated, it can be used in conjunction with the UCS of the intact rock σ_{ci} and the Hoek-Brown (HB) constant m_i and disturbance factor D to calculate the HB input parameters m_b, s and a. The GSI can also be used with some empirical models to estimate rock mass deformation and strength properties, as introduced in Chap. 1. This Chapter attempts to introduce the development of methods for estimating GSI value for rock masses.

J. Zuo and J. Shen, *The Hoek-Brown Failure criterion—From theory to application*,
https://doi.org/10.1007/978-981-15-1769-3_6

6.2 Background of GSI

Hoek (1994) noticed that a rock mass failure criterion would have no practical value unless it could be linked to geological description that could be made easily by engineering geologists. Although, the rock mass rating (RMR) (Bieniawski 1978) and Q classification system (Barton 1987) have been widely used in underground excavation and support, these two systems did not linked with rock mass strength and deformation properties which are usually used for the design and stability analysis in slope engineering.

The GSI classification started life in Toronto with engineering geology input from David Wood (Hoek et al. 1992) and were further developed by Hoek (1994), Hoek et al. (1995) and Hoek and Brown (1997). The value of GSI, which ranges from about 10 for extremely poor rock masses to 100 for intact rock, was roughly equivalent to RMR. The relations between GSI and RMR can be expressed as GSI = RMR_{89}-5. RMR_{89} is the 1989 version of RMR (Bieniawski 1989). The parameters used in RMR_{89} are listed in Table 6.1. The parameter J_{Cond} in Table 6.1 can be estimated from Table 6.2.

Hoek (1994) proposed a table that linked a geological description to an RMR value and to the HB input parameters m_b, s and a, which paved the way for

Table 6.1 Parameters used in RMR89 (Bieniawski 1989)

No.	Parameter	Rating
1	The strength of the intact rock material	0–15
2	Drill core quality (RQD)	3–20
3	Spacing of discontinuities	5–20
4	Condition of discontinuities (J_{Cond})	0–30
5	Groundwater	0–15
6	Adjustment for discontinuity orientation	−60 to 0

Table 6.2 Components of JCond89 (Bieniawski 1989)

Length	<1 m	1–3 m	3–10 m	10–20 m	>20 m
Rating	6	4	2	1	0
Separation	None	<0.1 mm	0.1–1 mm	1–5 mm	>5 mm
Rating	6	5	4	1	0
Roughness	Very rough	Rough	Slightly rough	Smooth	Slickensided
Rating	6	6	6	1	0
Infilling	None	Hard <5 mm	Hard >5 mm	Soft <5 mm	Soft >5 mm
Rating	6	4	2	2	0
Weathering	Unweathered	Slightly	Moderately	Highly	Decomposed
Rating	6	5	3	1	0

GENERALISED HOEK-BROWN CRITERION

$$\sigma_1' = \sigma_3' + \sigma_c \left[m_b \frac{\sigma_3'}{\sigma_c} + s \right]^a$$

σ_1' = major principal effective stress at failure

σ_3' = minor principal effective stress at failure

σ_c = uniaxial compressive strength of intact pieces of rock

m_b, s and a are constants which depend on the composition, structure and surface conditions of the rock mass

STRUCTURE	SURFACE CONDITIONS	VERY GOOD Very rough, fresh unweathered surfaces	GOOD Rough, slightly weathered, iron stained surfaces	FAIR Smooth, moderately weathered and altered surfaces	POOR Slickensided, highly weathered surfaces with compact coatings or fillings or angular fragments	VERY POOR Slickensided, highly weathered surfaces with soft clay coatings or fillings
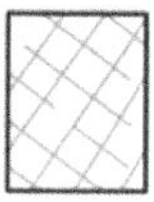 BLOCKY - well interlocked un - disturbed rock mass consisting of cubical blocks formed by three intersecting discontinuity sets	m_b/m_i	0.60	0.40	0.26	0.16	0.08
	s	0.190	0.062	0.015	0.003	0.0004
	a	0.5	0.5	0.5	0.5	0.5
	E_m	75,000	40,000	20,000	9,000	3,000
	ν	0.2	0.2	0.25	0.25	0.25
	GSI	85	75	62	48	34
 VERY BLOCKY - interlocked, partially disturbed mass with multi-faceted angular blocks formed by 4 or more joints sets	m_b/m_i	0.40	0.29	0.16	0.11	0.07
	s	0.062	0.021	0.003	0.001	0
	a	0.5	0.5	0.5	0.5	0.53
	E_m	40,000	24,000	9,000	5,000	2,500
	ν	0.2	0.25	0.25	0.25	0.3
	GSI	75	65	48	38	25
 BLOCKY/DISTURBED/SEAMY -folded with angular blocks formed by many intersecting discontinuity sets. Persistence of bedding planes or schistosity	m_b/m_i	0.24	0.17	0.12	0.08	0.06
	s	0.012	0.004	0.001	0	0
	a	0.5	0.5	0.5	0.5	0.55
	E_m	18,000	10,000	6,000	3,000	2,000
	ν	0.25	0.25	0.25	0.3	0.3
	GSI	60	50	40	30	20
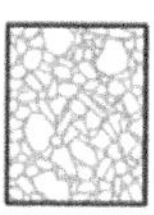 CRUSHED - poorly interlocked, heavily broken rock mass with mixture of angular and rounded blocks	m_b/m_i	0.17	0.12	0.08	0.06	0.04
	s	0.004	0.001	0	0	0
	a	0.5	0.5	0.5	0.55	0.60
	E_m	10,000	6,000	3,000	2,000	1,000
	ν	0.25	0.25	0.3	0.3	0.3
	GSI	50	40	30	20	10

Fig. 6.1 Chart linking geological description, GSI and the Hoek-Brown criterion (Hoek 1994)

assessing GSI directly from a chart (as shown in Fig. 6.1) with rock structures and surface condition of discontinuities rather than quantifying through RMR classification system.

6.3 The General GSI Chart

In the early days the use of the RMR system worked well because the majority of the engineering project were in reasonable quality rock masses (30 < RMR < 70) under moderate stress conditions. As the RMR classification is mainly dependent upon the value of RQD which is essentially equal to zero for the weak rock masses. Therefore, the RMR system was difficult to apply to rock masses that are of very poor quality, which means the correlations between RMR and the constants m_b and s of the HB criterion will be not reasonable for highly fractured rock masses. Also, Hoek (1994) stated that the orientation of discontinuity and groundwater parameters in RMR are dealt with explicitly in effective stress numerical analyses, thus the incorporation of these two parameters into the rock mass property is inappropriate. Therefore, it became necessary to consider to proposed an alternative rock mass classification system.

Based on dealing with incredibly difficult rock mass materials encountered in tunneling in Greece, Hoek et al. (1998), Marinos and Hoek (2000, 2001) proposed the general GSI system to include poor quality rock masses, as shown in Fig. 6.2. The general GSI classification system emphasis on basic geological observations of rock mass characteristics and was developed specifically for the estimation of strength and deformation properties of rock masses.

The heart of the GSI system is a careful engineering geology description of the rock mass encountered in engineering projects. The value of GSI is based upon the assessment of the two fundamental factors, namely, structure and discontinuities condition in the rock mass which can be estimated from visual examination of the rock mass exposed in outcrops, such as in excavated slope, tunnel faces and road cuts. Borehole cores are also the most reliable of data sources to estimate the value of GSI of the rock mass at certain depth, but it should be noted that multiple boreholes and inclined boreholes should be investigated to get reliable information of rock mass characteristics.

It should also be noted that the axes of GSI chart are not independent. The rock mass structure (vertical axis) is usually related to the surface condition (horizontal axis). This is commonly seen as a diagonal trend from top left to bottom right when various classes of the same type of rock masses are overlain on one chart (Bertuzzi et al. 2016). This kind of trend is observed in the charts published by Marinos and Hoek (2000) and Marinos et al. (2005) who noted that the top right and bottom left cells combinations are not available in the GSI chart shown in Fig. 6.2.

Currently, there are many GSI chart available for specific types of rock masses (Marinos and Hoek 2001; Marinos 2010; Hoek et al. 2013; Marinos et al. 2017; Marinos and Carter 2018). Each of the individual GSI charts describes the particular rock mass type in greater detail than the general chart. For example, Fig. 6.3 is a GSI chart for heterogeneous rock masses proposed by Marinos and Hoek (2001) and Marinos (2017). If the filling material in the discontinuities is systematic and thick (e.g. more than few cm) or shear zones are present with clayey material then the use of the heterogeneous GSI chart (Fig. 6.3) is recommended.

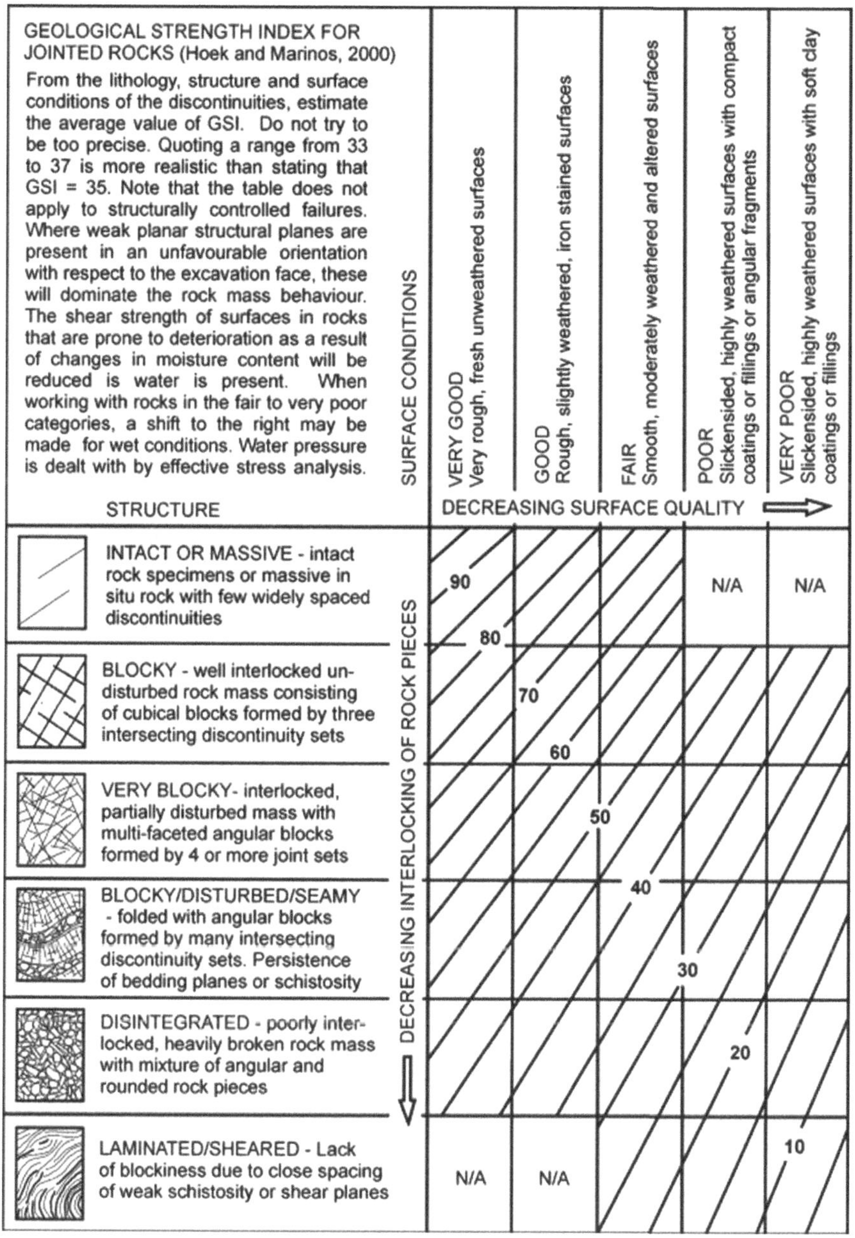

Fig. 6.2 The general GSI chart based on geological observations (Hoek and Marinos 2000)

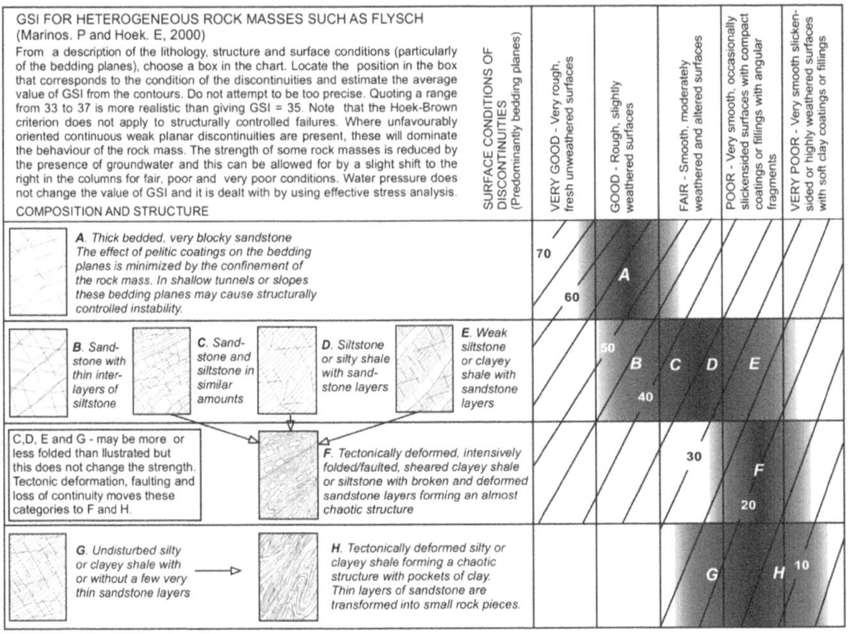

Fig. 6.3 Geological for heterogeneous rock masses (Marinos 2017)

Therefore, it is suggested that the selection of the appropriate GSI chart with visual similarity with the sketches of the structure of the rock mass domain when dealing with specific project. The rock mass characteristics should be descripted cautiously for the selection of the most suitable GSI chart (Marinos and Carter 2018).

6.4 Quantification of GSI Charts

Due to the lack of measurable and more representative parameters, and related interval limits or ratings for describing rock structure and the surface conditions of the discontinuities, the use of the general GSI chart involves some subjectivity. Therefore, several authors (Sonmez and Ulusay 1999; Cai et al. 2004; Hoek et al. 2013) proposed quantitative GSI charts to quantify the estimation of GSI to facilitate the use of the system especially by inexperienced engineers.

6.4.1 GSI Charts by Sonmez and Ulusay (1999)

The lack of parameters to describe the rock mass structure and surface conditions of the discontinuities prevents to obtain a more precise GSI value. For this reason, Sonmez and Ulusay (1999) made an attempt to provide a more quantitative basis for evaluating the GSI. Two factors, namely, structure rating (SR) and surface condition rating (SCR), were introduced into the general GSI chart. In order to assign the ratings to these two factors, the use of some easily measurable input parameters. They suggested that the value of SCR can be estimated from roughness, weathering and infilling of joints, as shown in Table 6.3.

The total rating for SCR is obtained using the following expression:

$$SCR = R_r + R_w + R_f \tag{6.1}$$

where R_r, R_w and R_f denote the ratings for roughness, weathering and infilling, respectively. The sum of the maximum ratings of these three parameters is 18.

Block size is an important index to quantify the rock mass structure. In the case of slopes, a small block size may cause rotational slides instead of structurally controlled modes of failure. In the case of tunneling, large blocks tend to be less deformable and develop favorable arching and interlocking in underground openings.

Block dimensions are determined by three rock mass parameters, namely the number of discontinuity sets, discontinuity spacing and the persistence of the discontinuities. However, in order to decrease the number of description input parameters, Sonmez and Ulusay (1999) used volumetric joint count J_v which is defined as the sum of the number of joints per meter for each joint set present, for the description of structure of the rock mass. J_v can be estimated from the following expressions:

$$J_v = \frac{N_1}{L_1} + \frac{N_2}{L_2} + \cdots + \frac{N_n}{L_n} \tag{6.2}$$

$$J_v = \frac{1}{S_1} + \frac{1}{S_2} + \cdots + \frac{1}{S_n} \tag{6.3}$$

Table 6.3 Components of surface condition rating (Bieniawski 1989)

Roughness	Very rough	Rough	Slightly rough	Smooth	Slickensided
Rating	6	6	6	1	0
Infilling	None	Hard <5 mm	Hard >5 mm	Soft <5 mm	Soft >5 mm
Rating	6	4	2	2	0
Weathering	Unweathered	Slightly	Moderately	Highly	Decomposed
Rating	6	5	3	1	0

where N is the number of joints along a scanline, L is the length of the scanline, S is the spacing, and n is the number of joint sets.

It should be noted that the estimation of J_v for heavily jointed rock masses with no identifiable structural pattern is not an easy task. Because the discontinuities in heavily jointed rock masses do not present extensive differences in their spacing in all directions. Sonmez and Ulusay (1999) suggested the following approach (Eq. 6.4) which is more practical in the estimation of the number of discontinuities in a rock mass with a volume of 1 m^3.

$$J_v = \frac{N_x}{L_x} + \frac{N_y}{L_y} + \frac{N_z}{L_z} \tag{6.4}$$

where N_x, N_y and N_z are the number of discontinuities counted along the scanlines, L_x, L_y and L_z are the length of the scanlines perpendicular to each other. However, when it is difficult to find out exposures along three scanline surveys in perpendicular directions, Eq. (6.5) can be used to estimate the value of J_v. In such circumstances, the rock mass is assumed as homogeneous.

$$J_v = \left(\frac{N}{L}\right)^3 \tag{6.5}$$

The relations between SR and J_v is shown in Fig. 6.4. The boundaries between the structural categories in the general GSI chart (Fig. 6.2) are equally divided.

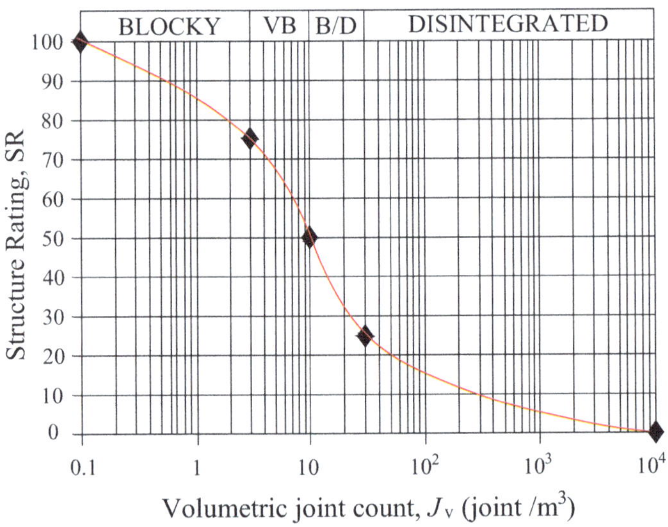

Fig. 6.4 The relations between structure rating and volumetric joint count (Sonmez and Ulusay 1999)

Table 6.4 Descriptive terms corresponding block size and intervals of J_v and GSI (ISRM 1981)

Descriptions J_v (joint/m^3)	Descriptions for GSI
Very large blocks <1	Blocky (B)
Large blocks 1–3	Blocky (B)
Medium sized blocks 3–10	Very Blocky (VB)
Small blocks 10–30	Blocky/Disturbed (B/D)
Very small blocks 30–60	Disintegrated (D)
Crushed >60	Disintegrated (D)

The SR limits between the B-VB, VBB/D and B/D-D rock structures are assumed as 75, 50 and 25, respectively.

The SR axis is divided into the ratings ranging from 0 to 100. Logarithmic J_v axis is divided according to the boundaries suggested for four structural categories (B, VB, B/D and D) as described in Table 6.4.

After the values of SR and SCR of a rock mass are estimated from Table 6.3 and Fig. 6.4, a quantitative GSI value can be estimated from the GSI chart as shown in Fig. 6.5.

6.4.2 GSI Charts by Cai et al. (2004)

Cai et al. (2004) presented a quantitative chart, as shown in Fig. 6.6, for evaluating the value of GSI from two indicators, namely, block volume (V_b) and joint condition factor (J_c).

Block size, which is determined from the joint spacing, number of joint sets, joint orientation and joint persistence, is an important index of rock mass quality. The block size can be calculated from the following equation.

$$V_b = \frac{s_1 s_2 s_3}{\sin \gamma_1 \sin \gamma_2 \sin \gamma_3} \tag{6.6}$$

where s_i and γ_i and are the joint spacing and the angle between joint sets as shown in Fig. 6.7.

When irregular jointing is encountered, it is not easy to delineate three or more joint sets. In such situation, the block volume can be directly measured in the field by selecting some representative blocks and measuring the irrelevant dimensions.

If the joints are not persistent, the equivalent block volume is expressed as:

$$V_b = \frac{s_1 s_2 s_3}{\sqrt[3]{p_1 p_2 p_3} \sin \gamma_1 \sin \gamma_2 \sin \gamma_3} \tag{6.7}$$

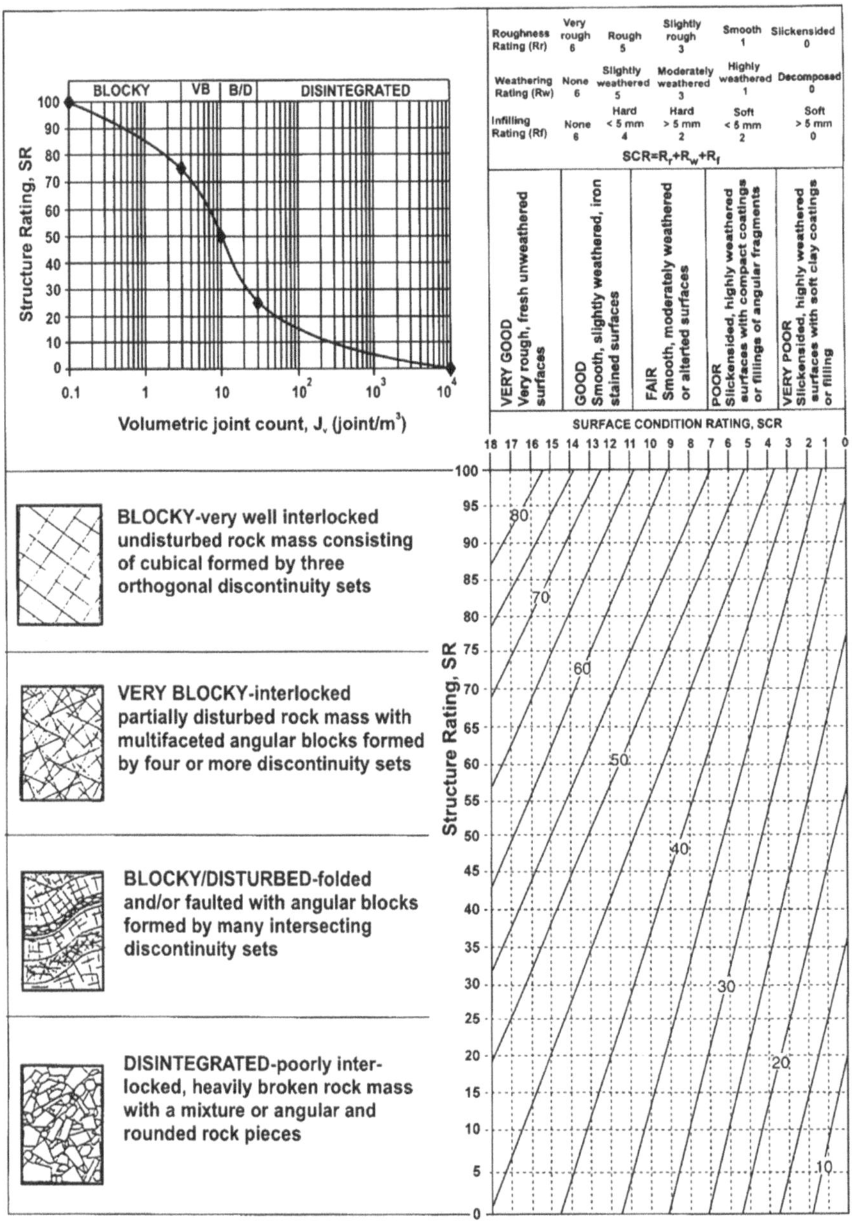

Fig. 6.5 Quantification of GSI chart (Sonmez and Ulusay 1999)

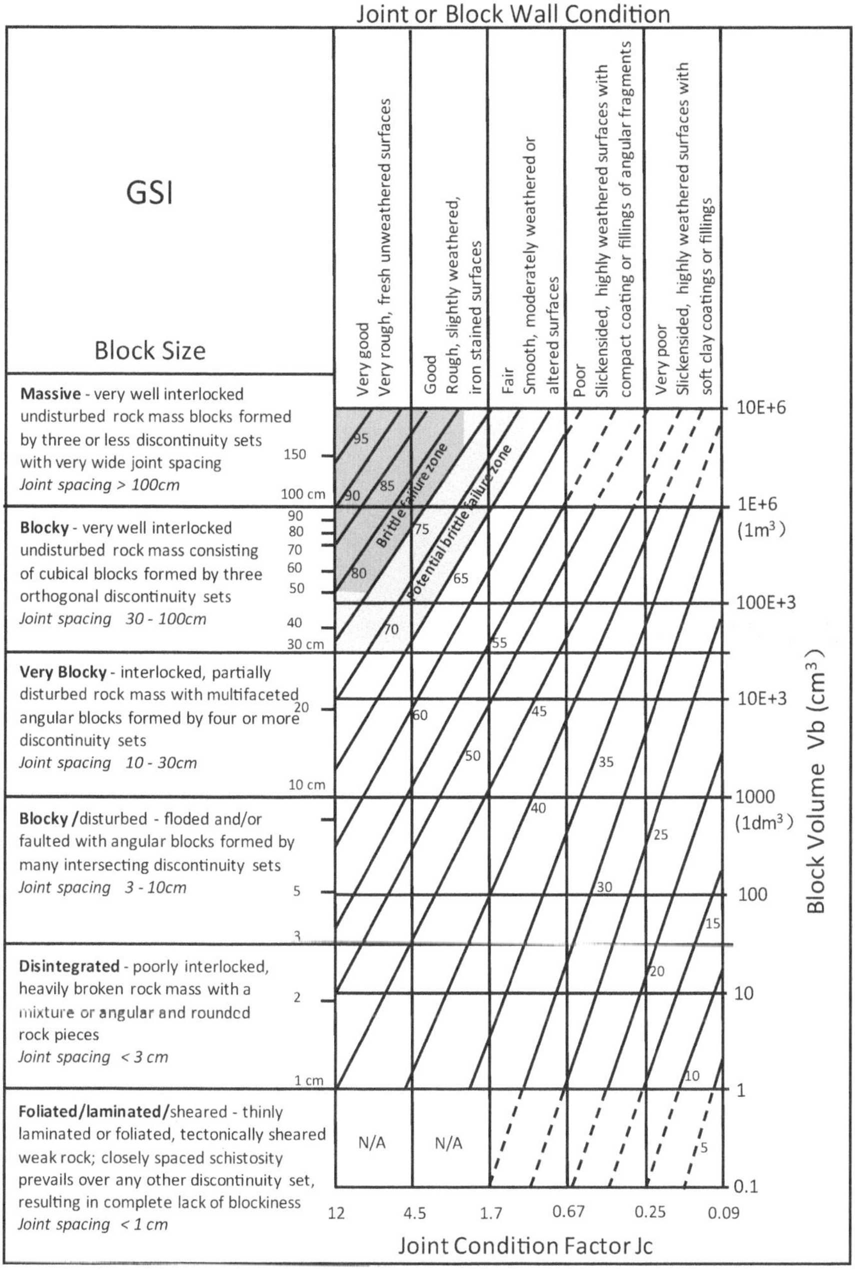

Fig. 6.6 Quantification of GSI chart (Cai et al. 2004)

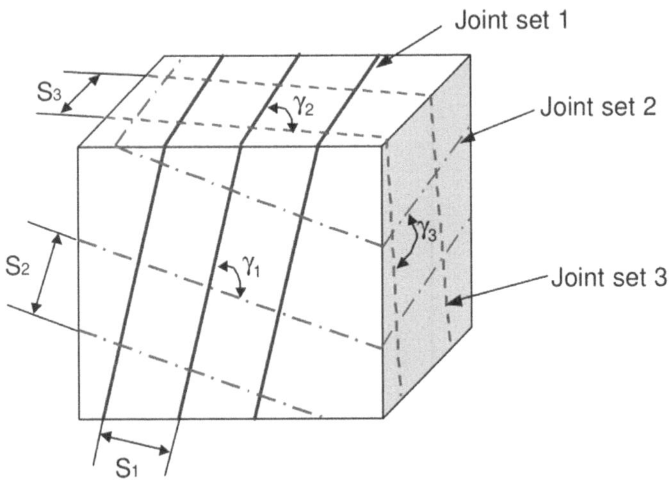

Fig. 6.7 Block delimited by three joint sets (Cai et al. 2004)

p_i is joint persistence factor which is defined as follows:

$$p_i = \frac{l_i}{L}\ when\ l_i < L \qquad (6.8)$$
$$p_i = 1\ when\ l_i \geq L$$

l_i is the accumulated joint length of set i in the sampling plane. L is the characteristic length of the rock mass.

On other hand, Cai et al. (2004) proposed a joint condition factor J_c to quantify the joint surface condition, which is defined as:

$$J_C = \frac{J_W J_S}{J_A} \qquad (6.9)$$

where J_W is the large-scale waviness (from 1 to 10 m) and can be estimated from Table 6.5. J_S is small-scale smoothness (from 1 to 20 cm) and can be estimated from Table 6.6. J_A is the joint alteration factor and can be estimated from Table 6.7.

Once the block volume (V_b) and the joint condition factor (J_c) are known, the GSI value can also be determined from the following equation proposed by Cai and Kaiser (2006)

$$\mathrm{GSI} = \frac{26.5 + 8.79 \ln J_c + 0.9 \ln V_b}{1 + 0.0151 \ln J_c - 0.0253 \ln V_b} \qquad (6.10)$$

Table 6.5 Terms to describe large-scale waviness J_W (Palmström 1995)

Waviness terms	Undulation *a/D*	Rating for waviness J_W
Interlocking (large-scale)		3
Stepped		2.5
Large undulation	>3%	2
Small to moderate undulation	0.3–3%	1.5
Planar	<0.3%	1

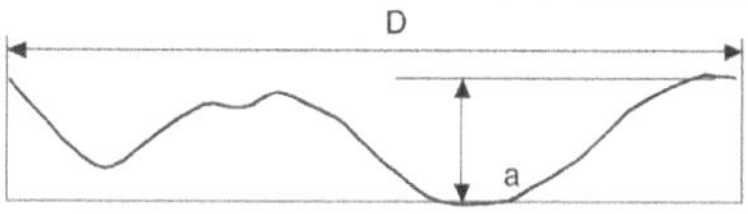

Undulation = a/D
D - length between maximum amplitudes

Table 6.6 Terms to describe small-scale smoothness J_S (Palmström 1995)

Smoothness terms	Description	Rating for smoothness J_S
Very rough	Near vertical steps and ridges occur with interlocking effect on the joint surface	3
Rough	Some ridge and side-angle are evident; asperities are clearly visible; discontinuity surface feels very abrasive (rougher than sandpaper grade 30)	2
Slightly rough	Asperities on the discontinuity surfaces are distinguishable and can be felt (like sandpaper grade 30–300)	1.5
Smooth	Surface appear smooth and feels so to touch (smoother than sandpaper grade 300)	1
Polished	Visual evidence of polishing exists. This is often seen in coating of chlorite and specially talc	0.75
Slickensided	Polished and striated surface that results from sliding along a fault surface or other movement surface	0.6–1.5

6.4.3 GSI Charts by Hoek et al. (2013)

Hoek et al. (2013) proposed a chart to quantify the GSI using the rock quality designation (RQD) and the joint condition rating (J_{Cond89}), as shown in Fig. 6.8 which set the vertical axis to RQD/2 and horizontal axis to 1.5 J_{Cond89}.

It can be seen that the top and bottom rows are removed compared with the general GSI chart in Fig. 6.2. Therefore, the applicable range for GSI is limited to RQD $\leq$ 80. Once the RQD and the J_{Cond89} are estimated, the GSI value can also be determined from the following equation:

Table 6.7 Rating for the joint alteration factor JA (Barton 1974; Palmström 1995)

	Term	Description	J_A
Rock wall contact	Clear joints healed or "welded" joints (unweathered)	Softening, impermeable filling (quartz, epidote, etc.)	0.75
	Fresh rock walls (unweathered)	No coating or filling on joint surface, except for staining	1
	Alteration of joint wall: slightly to moderately weathered	The joint surface exhibits one class higher alteration than the rock	4
	Alteration of joint wall: highly weathered	The joint surface exhibits two classes higher alteration than the rock	4
	Coating or thin filling Sand, silt, calcite, etc.	Coating of frictional material without clay	3
	Clay, chlorite, talc, etc.	Coating of softening and cohesive minerals	4
	Sand, silt, calcite, etc.	Filling of frictional material without clay	4
Filled joints with partial or no contact between the rock wall surfaces	Compacted clay materials	"Hard" filling of softening and cohesive materials	6
	Soft clay materials	Medium to low over-consolidation of filling	8
	Swelling clay materials	Filling material exhibits swelling properties	8–12

$$\mathrm{GSI} = 1.5 \times Jcond_{89} + RQD/2 \tag{6.11}$$

In recognition of the fact that values of J_{Cond89} are not always available in data from field mapping, Hoek et al. (2013) suggested that the version of Joint Condition rating (J_{Cond78}), as shown in Table 6.8, can be used to calculate J_{Cond89}. Regression analysis of a plot of individual values assigned to J_{Cond78} and J_{Cond89} gives $J_{Cond89} = 1.3\ J_{Cond78}$.

The values of J_{Cond89} can also be estimated from joint roughness number J_r and joint alteration number J_a. The values of J_r and J_a can be estimated from Table 6.9. The quotient J_r/J_a which is included in the tunnelling quality index proposed by Barton et al. (1974), which represents the roughness and frictional characteristics of the joint walls or fillings.

Comparing the ratings for $J_{Cond\ 89}$ with those allocated to J_r and J_a gives the relation as follows:

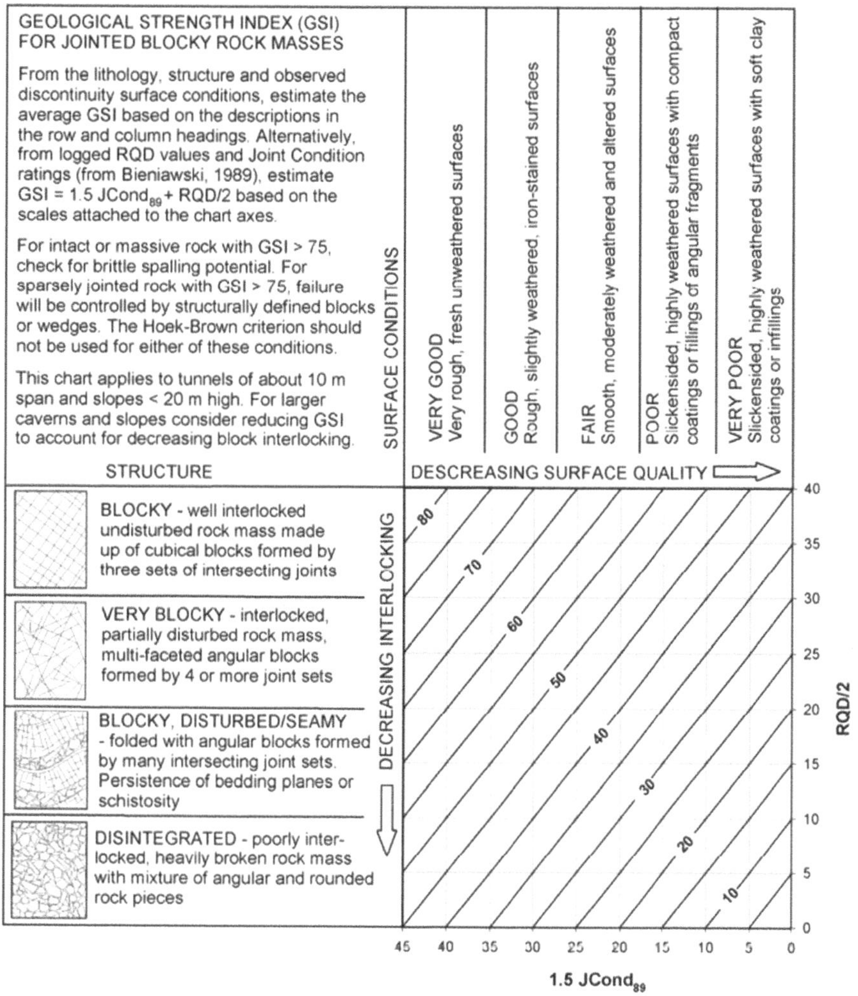

Fig. 6.8 Quantification of GSI chart (Hoek et al. 2013)

$$J_{Cond89} = \frac{35J_r/J_a}{(1 + J_r/J_a)} \tag{6.12}$$

RQD was developed by Deere et al. (1968) and can provide a quantitative estimate of rock mass quality from drill core logs. Therefore, it was adopted to quantify the rock mass structure by Hoek et al. (2013). RQD is defined as the percentage of intact core pieces longer than 100 mm in the total length of core (measured along core centreline), as follows:

Table 6.8 Definition of J_{Cond78} (Bieniawski 1978)

Condition of discontinuities	Very rough surfaces Not continuous No separation Hard joint wall rock	Slightly rough surfaces Separation <1 mm Hard joint wall rock	Slightly rough surfaces Separation <1 mm Soft joint wall rock	Slickensided surfaces or Gouge <5 mm thick or Joints open 1–5 mm Continuous joints	Soft gouge >5 mm thick or Joints open >5 mm Continuous joints
Rating	25	20	12	6	0

Table 6.9 Definition of J_r and J_a for rock wall contact (no pre-shearing) (Barton et al. 1974)

Joint roughness number J_r	Rating	Joint alteration number J_a	Rating
Discontinuous joints	4	Tightly healed, hard, non-softening, impermeable filling	0.75
Rough and irregular, undulating	3	Unaltered joint walls, surface staining only	1.0
Smooth, undulating	2	Slightly altered joint walls, non-softening mineral coatings, sandy particles, clay-free disintegrated rock, etc.	2.0
Slickensided, undulating	1.5	Silty-, or sandy-clay coatings, small clay fraction (non-softening)	3.0
Rough or irregular planar	1.5	Softening or low friction clay, mineral coatings, i.e. kaolinite, mica. Also chlorite, talc, gypsum and graphite etc., and small quantities of swelling clays. (Discontinuous coatings, 1–2 mm or less in thickness)	4.0
Smooth, planar	1.0		
Slickensided, planar	0.5		

$$RQD = \frac{\sum \text{length of core pieces}}{\text{Total length of core sun}} \times 100 \tag{6.13}$$

However, RQD is a directionally dependent parameter and its value may change significantly with the changes of borehole orientation. Priest and Hudson (1976) suggested that the value of RQD could also be obtained from discontinuity spacing measurements made on core or from an exposure, as follows:

$$RQD = 100e^{0.1\lambda}(0.1\lambda + 1) \tag{6.14}$$

where λ is the average number of discontinuities per meter.

6.5 When Not to Use GSI

Figure 6.9 is the typical HB idealized diagram showing the transition from intact rocks to increasing heavily jointed rock mass, which can be used to explain the range of applicability of the GSI system. It shows that the GSI should be used in cases for intact rocks or heavily jointed rock masses and should not be used in cases where there are only one or two sets of discontinuities. Great caution should be taken when apply for blocky rock mass with the minimal anisotropy. However, in a numerical rock mass model containing a single well-defined discontinuity, it is sometimes reasonable to assign the GSI to the overall rock mass and to add the single discontinuity as a layer element in the numerical model (Hoek and Brown 2018).

Fig. 6.9 Definition of RQD (after Deere 1963)

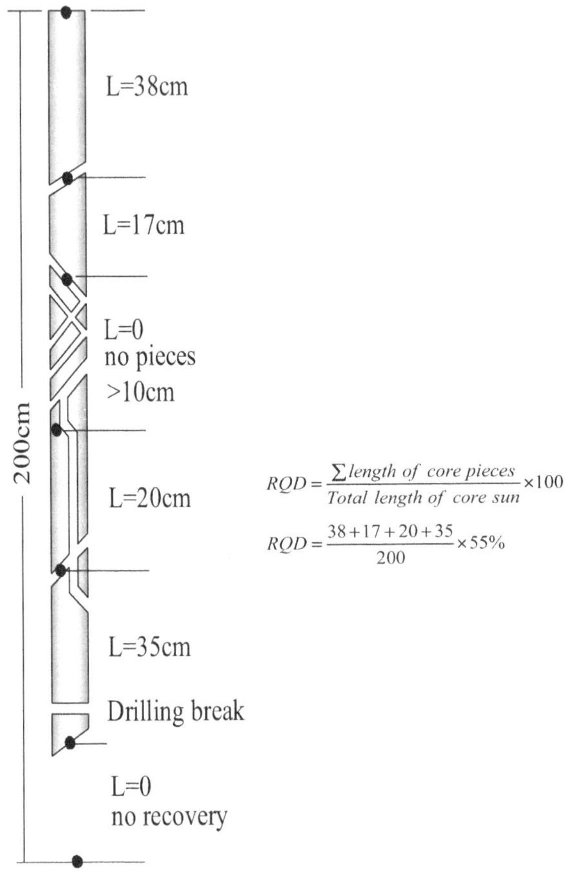

On the other hand, a fundamental assumption of the HB criterion for the estimation of the strength and deformation of rock masses is that rock mass deformation and failure are controlled by sliding and rotation of intact blocks of rock defined by intersecting discontinuity systems. Therefore, for deep mining engineering, such as depth more than 1000 m, the rock mass structure is so tight under the high stress condition and the mass behavior approaches that of the intact rock. The failure process of rock masses that controls the stability of underground excavations under these high stress conditions is dominated by brittle fracture initiation and propagation, which would lead to spalling, slabbing and rock-bursts. In such situation, the value of GSI tends to be equal to 100 and the application of the GSI system to estimate rock mass deformation and strength of rock masses is no longer meaningful (Hoek and Brown 2018) (Fig. 6.10).

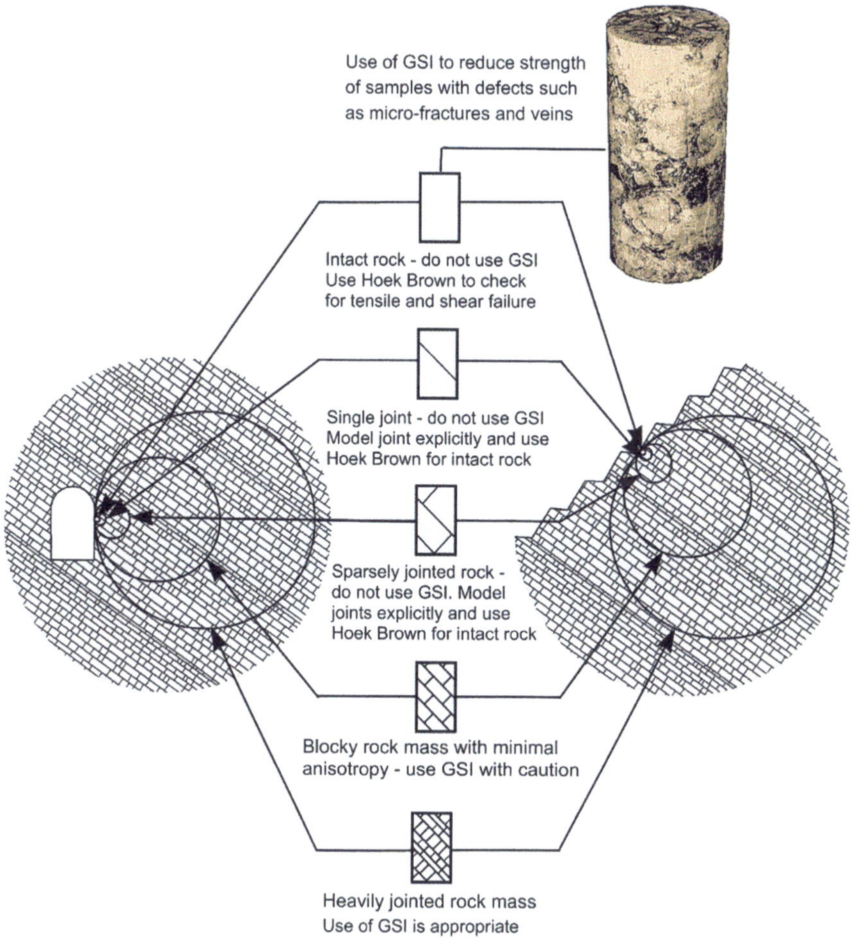

Fig. 6.10 Idealised diagram showing the transition from intact to a heavily joined rock mass with increasing sample size (Hoek and Brown 2018)

References

Barton NR, Lien R, Lunde J (1974) Engineering classification of rock masses for the design of tunnel support. Rock Mech 6(4):189–239

Barton N (1987) Rock mass classification, tunnel reinforcement selection using the Q-system. In: Proceedings of the ASTM symposium on rock classification systems for engineering purposes. Cincinnati, Ohio

Bieniawski ZT (1978) Determining rock mass deformability e experience from case histories. Int J Rock Mech Min Sci Geomech Abstr 15(5):237–247

Bieniawski ZT (1989) Engineering rock mass classifications: a complete manual for engineers and geologists in mining, civil and petroleum engineering. Wiley, New York

Bertuzzi R, Douglas K, Mostyn G (2016) Comparison of quantified and chartGSIfor four rock masses. Eng Geol 202:24–35
Cai M, Kaiser P (2006) Visualization of rock mass classification systems. Geotech Geol Eng 24 (4):1089–1102
Cai M, Kaiser P, Uno H, Tasaka Y, Minami M (2004) Estimation of rock mass strength and deformation modulus of jointed hard rock masses using the GSI system. Int J Rock Mech Min Sci 41(1):3–19
Deere DU (1963) Technical description of rock cores for engineering purposes. Rock Mech Eng Geol 1(1):16–22
Deere D (1968) Geological considerations. In: Stagg KG, Zienkiewicz OC (eds) Rock mechanics in engineering practice. Wiley, London, pp 1–20
Hoek E (1994) Strength of rock and rock masses. News J ISRM 2(2):4–16
Hoek E, Brown E (1997) Practical estimates of rock mass strength. Int J Rock Mech Min Sci Geomech Abstr 34:1165–1186
Hoek E, Brown E (2018) The Hoeke-Brown failure criterion and GSI-2018 edition. J Rock Mech Geotech Eng. https://doi.org/10.1016/j.jrmge.2018.08.001
Hoek E, Carter T, Diederichs M (2013) Diederichs quantification of the geological strength index chart 47th US rock mechanics/geomechanics symposium (ARMA 13-672, San Francisco, CA, USA)
Hoek E, Kaiser P, Bawden W (1995) Support of underground excavations in hard rock. AA Balkema, Rotterdam
Hoek E, Marinos P, Benissi M (1998) Applicability of the geological strength index (GSI) classification for weak and sheared rock masses—the case of the Athens schist formation. Bull Eng Geol Env 57(2):151–160
Hoek E, Wood D, Shah S (1992) A modified Hoek–Brown criterion for jointed rock masses. In: Hudson JA (ed) Proceedings of the rock mechanic symposium. International society of rock mechanics Eurock '92, British geotechnical society, London, pp 209–214
ISRM (International Society for Rock Mechanics) (1981) In: Brown ET (ed) ISRM suggested methods: rock characterization, testing and monitoring. Pergamon Press, London
Marinos P (2010) New proposed GSI classification charts for weak or for complex rock masses 12th international congress. Bulletin of the Geological Society of Greece, Patras, pp 1248–1258
Marinos V (2017) A revised geotechnical classification GSI system for tectonically disturbed rock masses, such as flysch. Bull Eng Geol Env 19:1–14
Marinos V, Carter TG (2018) Maintaining geological reality in application of GSI for design of engineering structures in rock. J Eng Geol 239:282–297
Marinos P, Hoek E (2000) GSI: a geologically friendly tool for rock mass strength estimation. In: Proceedings of the GeoEng2000 at the international conference on geotechnical and geological engineering, Melbourne, Technomic Publishers, Lancaster, pp 1422–1446
Marinos P, Hoek E (2001) Estimating the geotechnical properties of heterogeneous rock masses such as flysch. Bull Eng Geol Environ 60:82–92
Marinos V, Marinos P, Hoek E (2005) The geological strength index: applications and limitations. Bull Eng Geol Env 64(1):55–65
Palmström A (1995) RMi—a rock mass characterization system for rock engineering purposes. PhD thesis, University of Oslo, Norway
Priest S, Hudson J (1976) Discontinuity spacings in rock Int J Rock Mech Min Sci Geomech Abstr 13(5):135–148
Sonmez H, Ulusay R (1999) Modifications to the geological strength index (GSI) and their applicability to the stability of slopes. Int J Rock Mech Min Sci 36:743–760

Chapter 7
The Blast Damage Factor *D*

Abstract Practical experience in the design of rock slope projects has demonstrated that the estimated rock mass strength of a given slope is affected by the blast damage in slope excavations. In order to improve the accuracy of rock mass strength prediction under blast conditions, a blast damage factor D, also known as the disturbance factor, was introduced in the Hoek-Brown criterion to calculate the HB input parameters and strength and deformation modulus of rock masses. The value of D varies from zero for undisturbed rock masses to one for highly disturbed rock masses according to disturbance conditions caused by blast damage as well as stress relief. Guidelines for estimating disturbance factor D for the tunnel and slope application were presented. Some preliminary guidelines as a starting point for the determination of the thickness of the blast damage zone as a result of production blast were also presented.

7.1 Introduction

Practical experience in the design of rock slope projects has demonstrated that the estimated rock mass strength of a given slope is affected by the blast damage in slope excavations (Hoek and Brown 1988). In order to improve the accuracy of rock mass strength prediction under blast conditions, a blast damage factor D, also known as the disturbance factor, was introduced in the Hoek-Brown (HB) criterion (Eq. 7.1) to calculate the HB input parameters (Hoek et al. 2002), as follows:

$$\sigma_1 = \sigma_3 + \sigma_{ci}\left(m_b \frac{\sigma_3}{\sigma_{ci}} + s\right)^a \tag{7.1}$$

where σ_1 and σ_3 are the maximum and minimum principal stresses at failure, respectively; σ_{ci} is the uniaxial compressive strength (UCS) of the intact rock. m_b, s and a are the HB input parameters which can be estimated from the HB constant m_i of intact rocks, the Geological Strength Index (GSI) and the blast damage factor D, given by:

J. Zuo and J. Shen, *The Hoek-Brown Failure criterion—From theory to application*,
https://doi.org/10.1007/978-981-15-1769-3_7

$$m_b = m_i \exp(\frac{GSI - 100}{28 - 14D}) \tag{7.2}$$

$$s = \exp(\frac{GSI - 100}{9 - 3D}) \tag{7.3}$$

$$a = \frac{1}{2} + \frac{1}{6}(e^{-GSI/15} - e^{-20/3}) \tag{7.4}$$

The value of m_i depends on rock types, mineral compositions, grain sizes and cementation of rocks, and it can be estimated based on regression analyses of triaxial experimental data of intact rocks. GSI depends on rock mass structures and surface conditions. The details of calculation and selection of m_i and GSI can be found in Chaps. 5 and 6. This Chapter will introduce the selection of the disturbance factor D and the disturbance zone thickness T for rock masses for slope and tunnel projects. The effect of D on rock mass properties is also discusses.

7.2 Guidelines for Estimating Disturbance Factor *D*

Engineering experience in the design of large open pit slopes has demonstrated that it is not easy to estimate the accurate value of D as various factors can influence the degree of disturbance in the rock mass. Hoek et al. (2002), Hoek and Diederichs (2006), Hoek (2007), Hoek and Brown (2018), therefore, presented a number of tunnel and slope cases to illustrate how to select reasonable values of D for practical projects. Guidelines for estimating disturbance factor D for the tunnel and slope application can be found in Table 7.1. The value of D varies from zero for undisturbed rock masses to one for highly disturbed rock masses according to disturbance conditions caused by blast damage as well as stress relief.

In the case of tunnels, this is particularly critical since the limited amount of space available in a tunnel means that any rock mass failure can have a significant impact on the excavation schedule. Careful excavation by road-header or a tunnel boring machine can reduce the disturbance effects on rock masses. However, when the tunnel is excavated by drill-and-blast techniques, the stability of rock masses will be significantly influenced by the blasting design (Hoek 2012).

Five typical cases were presented to illustrate how to assign appropriate D values to engineering projects under various excavation conditions (Hoek and Brown 2018).

The first scenario in Table 7.1 presents a tunnel in which the blast-hole pattern, explosive charges, and detonation sequence were carefully designed and executed. The tunnel walls were created by careful control of smooth blasting. In such case, the $D = 0$ can be assigned to the tunnel with slight damage to the surrounding rock mass.

Table 7.1 Guidelines for estimating disturbance factor D (Hoek 2012)

Rock mass appearance	Description of rock mass	Suggested value of D
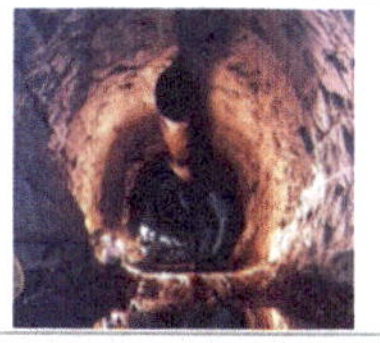	Excellent quality controlled blasting or excavation by Tunnel Boring Machine results in minimal disturbance to the confined rock mass surrounding a tunnel	$D = 0$
	Mechanical or hand excavation in poor quality rock masses (no blasting) results in minimal disturbance to the surrounding rock mass. Where squeezing problems result in significant floor heave, disturbance can be severe unless a temporary invert, as shown in the photograph, is placed	$D = 0$ $D = 0.5$ No invert
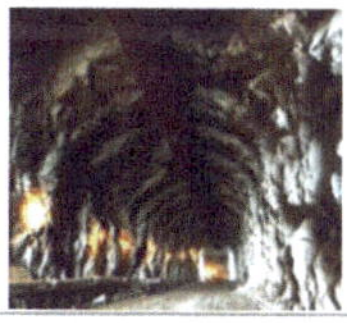	Very poor quality blasting in a hard rock tunnel results in severe local damage, extending 2 or 3 m, in the surrounding rock mass	$D = 0.8$
	Small scale blasting in civil engineering slopes results in modest rock mass damage, particularly if controlled blasting is used as shown on the left hand side of the photograph However, stress relief results in some disturbance	$D = 0.7$ Good blasting $D = 1.0$ Poor blasting
	Very large open pit mine slopes suffer significant disturbance due to heavy production blasting and also due to stress relief from overburden removal In some softer rocks excavation can be carried out by ripping and dozing and the degree of damage to the slopes is less	$D = 1.0$ Production blasting $D = 0.7$ Mechanical excavation

The second case is a more complex situation which shows a tunnel excavated by the top heading and bench method. Excessive displacements in the lower part of the tunnel can result in significant rock mass failure if the displacements induced by the excavation of the lower bench are not carefully controlled by the placement of an inverted strut. In such situation, $D = 0.5$ is assigned for the rock mass in which the lower half of the tunnel is excavated. It should be noted that the value of D should only be applied to the surrounding rock mass of about 2 m width around the bottom half of this 12 m span tunnel.

The third case is a very poorly designed and executed tunnel. Poor drill holes alignment control and lack of attention to the blast design and detonation sequence were resulted in heave damage to the surrounding rock mass. $D = 1.0$, with a linear decrease to zero, was reasonably assigned to the first 3 m of the rock mass surrounding this 8 m span tunnel.

The fourth case illustrates a 15 m high slope in which pre-split blasting method was used to create the face on the left. Relaxation of the face can still occur and $D = 0.5$ was assigned to 1–2 m of rock behind this face. The rock mass on the right was mass blasted with little control of the drill hole spacing and alignment of the charges and detonation sequence. The most severe disturbance factor value is D = 1.0 was assigned to 2–3 m of the rock mass behind this slope.

The fifth illustration in Table 7.1 shows a very large open pit with the height of 1000 m. Various D values should be considered in this case. It is important to identify the difference between slopes created during active mining excavation which is required to produce large tons of ore and the final design slopes which are required to remain stable to ensure access to the ore and safe and efficient disposal of the waste. The benches with 18 m height were suffered significant damage because of their proximity to the production blasts required for removal and fragmentation of the ore. $D = 1.0$ is assigned to the rock immediately behind these benches. The value of D can be graded downward to $D = 0$, as the distance behind the face increases to about 30% of the slope height.

7.3 Guidelines for Estimating Disturbance Zone Thickness *T*

The thickness of the disturbance zone (see Fig. 7.1) depends on blast design factors (explosive type, charge, drill hole diameter and burden) and rock mass mechanical properties. Lupogo (2017) noted that the persistence of pre-existing discontinuities could significantly influence the thickness of blast damage zone.

The rock mass with a lower discontinuity persistence tends to develop a low thickness of blast damage zone compared to a rock mass with a high discontinuity persistence. Furthermore, it was observed that the blast damage zone thickness decreases with an increase in the tensile strength of the rock mass. Mohanty and

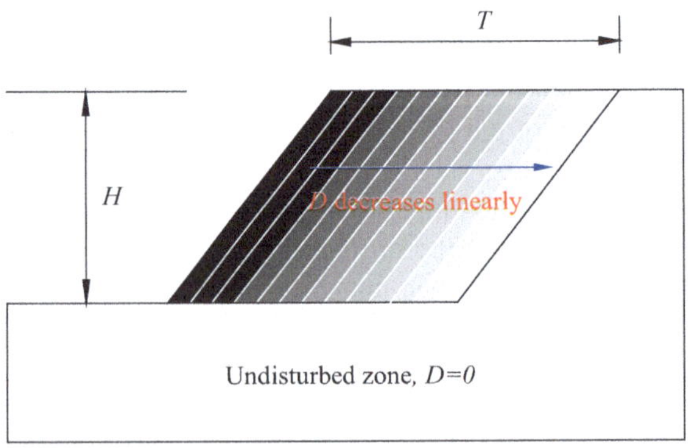

Fig. 7.1 Diagrammatic representation of slope models considering disturbance zone

Chung (1986) divided the blast damage zone into two types in open pit mines: near-field damage which is caused by high frequency vibration when the explosion occurs in the vicinity of the sloped wall (0–20 m), and far field (up to 100 m) damage which is caused by low frequency vibration. It should be noted that backbreaker, which can be defined as over break of rocks beyond the limits of the rear row of holes in a blast pattern, could be observed in the blast damage zone due to inappropriate blast design. This phenomenon can cause problems of mine wall instability and improper fragmentation, and, consequently, increase the cost of a mining operation. Hoek (2012) presented some preliminary guidelines as a starting point for the determination of the thickness of the blast damage zone as a result of production blast. Table 7.2 shows that for a large production blast, $T = 2.0$–$2.5H$ when the blast is confined with little or no control; $T = 1.0$–$1.5H$ when the blast is to a free face with no control; and $T = 0.5$–$1.0H$ when the blast is to a free face with some controls, e.g. one or more buffer rows used.

Zheng et al. (2018) used the parallel layer model (PLM), which rock mass is divided into a number of layers parallel to slope surface with a decreasing value of D applied to each layer in slope modeling, in the limit equilibrium method to investigate the effect of blast damage zone T on the stability of rock slopes. Results show that the selection of T in the slope model plays an important role in terms of calculation of FOS, especially, when the ratio of T to slope height H is less than 1.0.

7.4 Effect of *D* on Rock Mass Properties

It is not easy to determine the exact value of D as various factors can influence the degree of disturbance in the rock mass, such as the design of blasting and rock mass properties. Therefore, based on a number of case studies, Hoek et al. (2002) and Hoek (2012) provided some guidance on the selection of appropriate D values for practical application.

Table 7.2 Guidelines for estimating disturbance zone thickness T (Hoek 2012)

Blasting conditions	The suggested T values
Large production blast, confined with little or no control	$T = 2.0$–2.5H
Production blast with no control but blasting to a free face	$T = 1.0$–1.5H
Production blast, confined but with some control, e.g. one or more buffer rows	$T = 1.0$–1.2H
Production blast with some control, e.g. one or more buffer rows, and blasting to a free face	$T = 0.5$–1.0H
Carefully controlled production blast with a free face	$T = 0.3$–0.5H

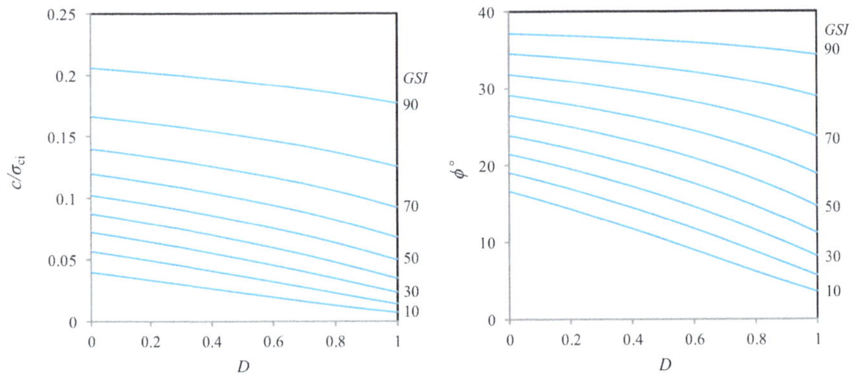

(a) The influence of D on c/σ_{ci} (b) The influence of D on ϕ

Fig. 7.2 The influence of the disturbance factor D on shear strength parameters when $\sigma_{ci}/\gamma H = 1$ and $m_i = 15$

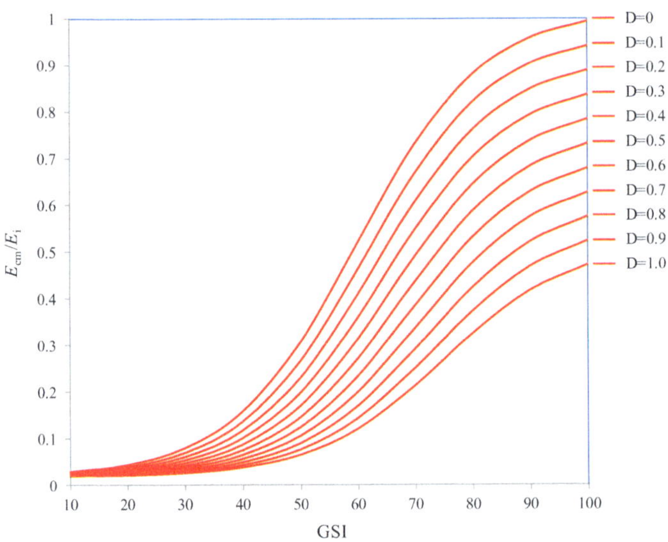

Fig. 7.3 The influence of the disturbance factor D on deformation modulus (after Hoek and Diederichs 2006)

While D values have great influence on the rock masses properties, it is, nevertheless, difficult to determine its exact value. Therefore, a sensitivity analysis of D is probably more important in judging the acceptability of a slope design than the results from a single D analysis. Such analysis was presented in Figs. 7.2, 7.3 and

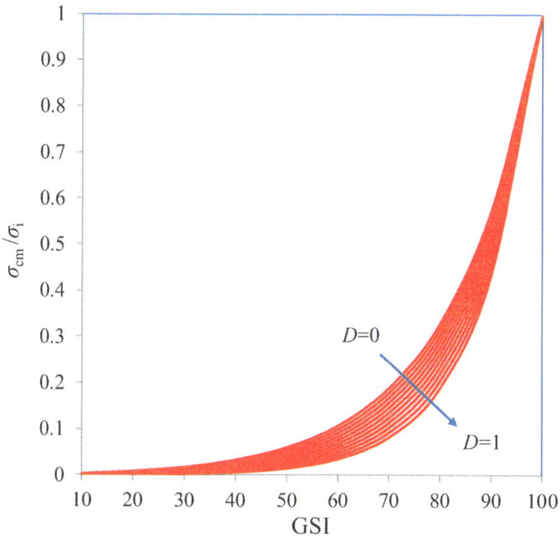

Fig. 7.4 The influence of the disturbance factor *D* on UCS of rock masses based on Eq. 1.8

7.4 which illustrates the influence of *D* on the rock mass shear strength parameters and deformation modulus and UCS of rock masses. *D* values range between 0 or undisturbed rock and 1 for highly disturbed rock mass.

Wan et al. (2016) proved that for a given slope or tunnel with known of the Hoek-Brown parameters GSI, m_i and *D*, the values of c/σ_{ci} and ϕ are depend on the dimensionless parameter SR which is the ratio of σ_{ci} to γH, regardless of the values of individual parameters σ_{ci}, γ and *H*.

Figures 7.5 and 7.6 presented charts which can be used to estimate the MC shear strength parameters of homogenous rock masses from the Hoek-Brown parameters (GSI, m_i and *D*), slope geometry (*H*) and rock mass properties (σ_{ci} and γ) for the design of rock slopes based on the Eqs. (1.27–1.32).

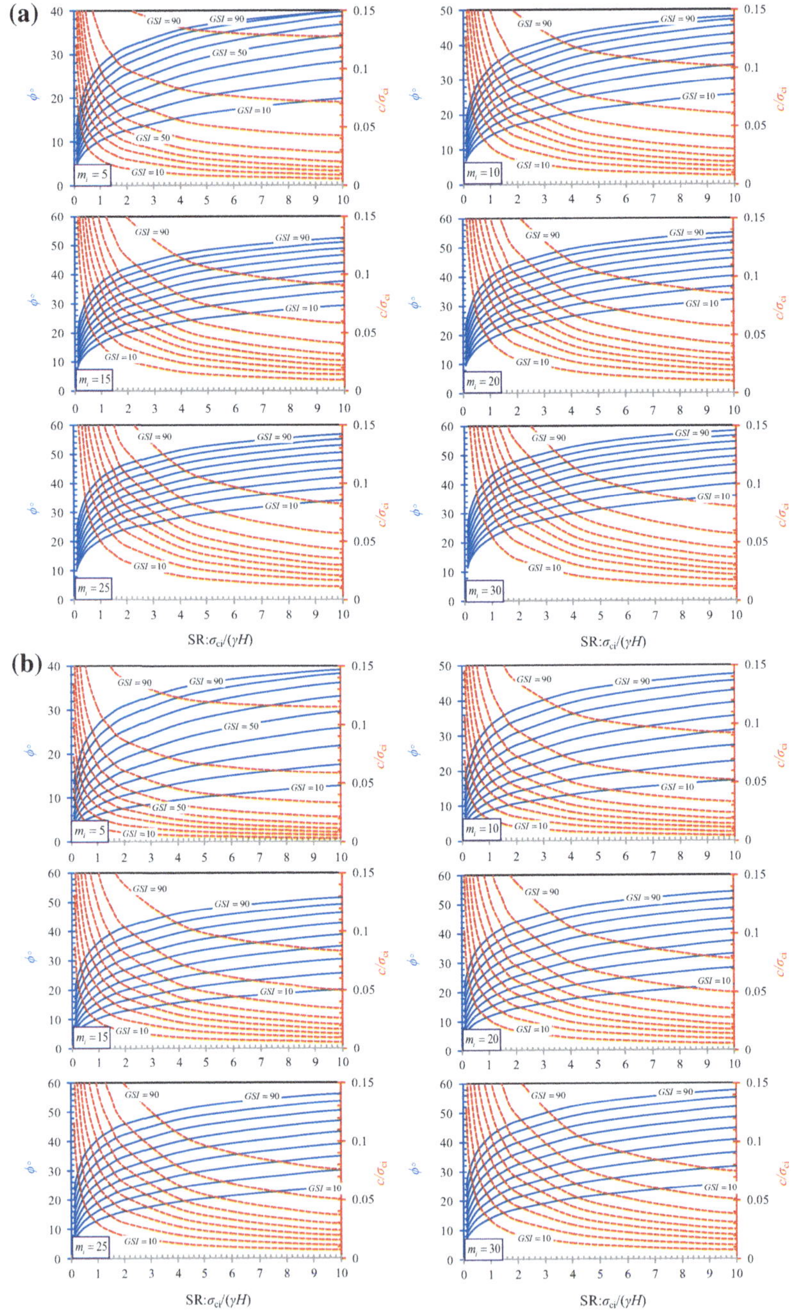

Fig. 7.5 Charts for estimating shear strength parameters for slopes. **a** D = 0, **b** D = 0.5, **c** D = 1.0

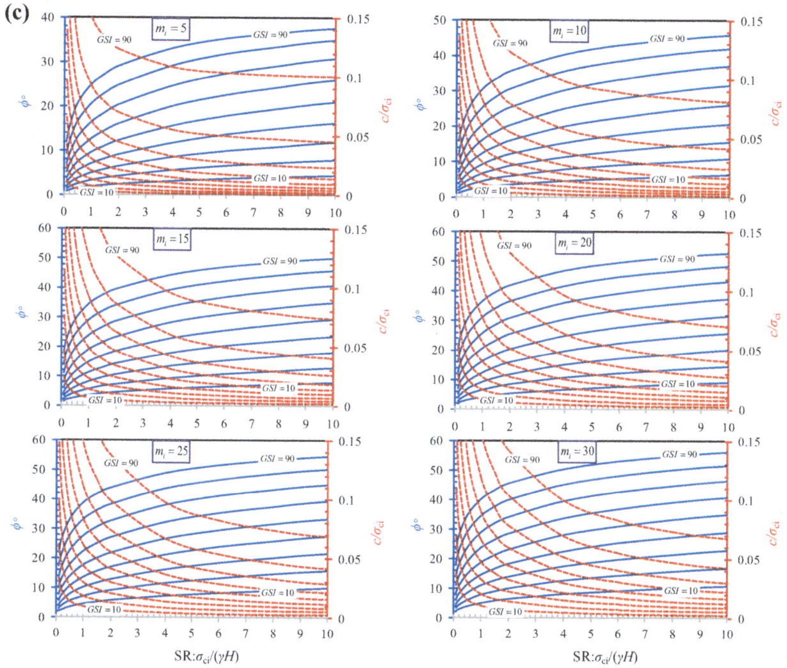

Fig. 7.5 (continued)

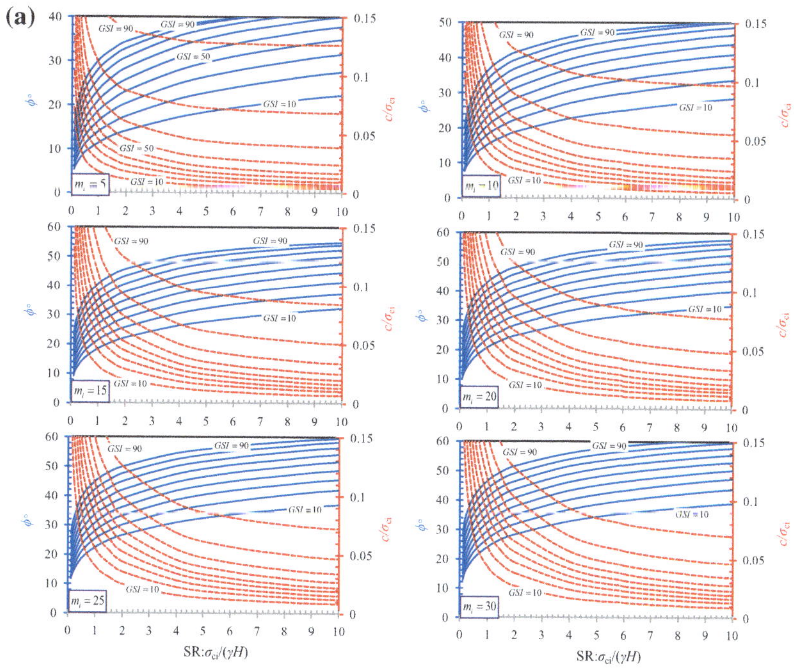

Fig. 7.6 Charts for estimating shear strength parameters for tunnels. **a** $D = 0$, **b** $D = 0.5$, **c** $D = 1.0$

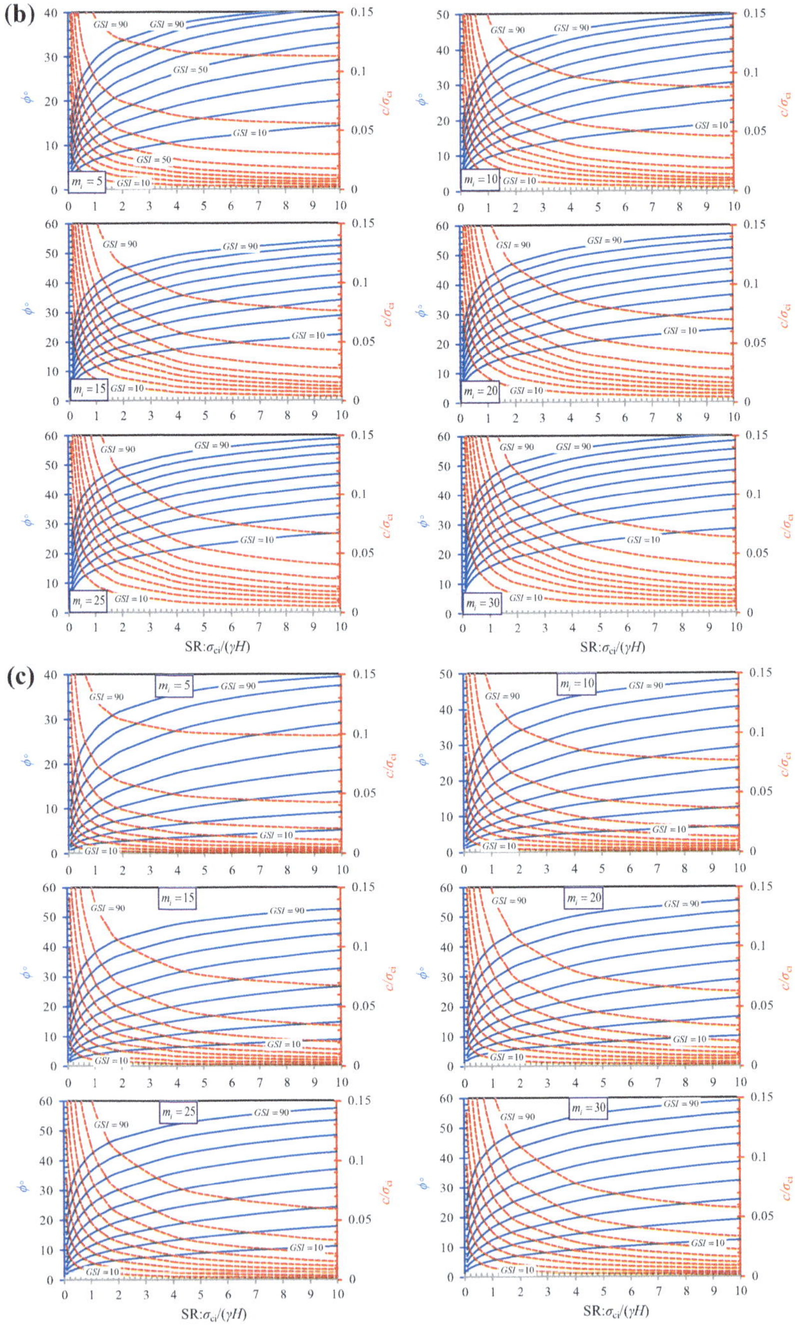

Fig. 7.6 (continued)

References

Hoek E (2007) Rock mass properties (Chapter 11) Practical rock engineering. https://www.rocscience.com/documents/hoek/corner/12_Rock_mass_properties.pdf. Accessed 20 Apr 2018

Hoek E (2012) Blast damage factor D. Technical note for RocNews. https://www.rocscience.com/documents/pdfs/rocnews/winter2012/Blast-Damage-Factor-D-Hoek.pdf. Accessed 20 Apr 2018

Hoek E, Brown ET (1988) The Hoek-Brown failure criterion—a 1988 update. In: Curran JC (eds) Proceedings of 15th Canadian rock mechanics symposium, pp 31–38

Hoek E, Brown ET (2018) The Hoek-Brown failure criterion and GSI. J Rock Mech Geotech Eng. https://doi.org/10.1016/j.jrmge.2018.08.001 (2018 edition)

Hoek E, Carranza-Torres C, and Corkum B (2002) Hoek-Brown failure criterion. In: Hammah R, Bawden W, Curran J, Telesnicki M (eds) Proceedings of NARMS-TAC 2002, mining innovation and technology, Toronto, Ontario (2002 edn)

Hoek E, Diederichs MS (2006) Empirical estimation of rock mass modulus. Int J Rock Mech Min Sci 43(2):203–215

Lupogo K (2017) Characterization of blast damage in rock slopes: an integrated field-numerical modeling approach. PhD thesis, Simon Fraser University, British Columbia, Canada

Mohanty B, Chung S (1986) Production blasting and slope stability. In: Proceedings of the international symposium on geotechnical stability in surface mining, CANMET, Calgary, Alberta, Canada, pp 133–140

Wan L, Wei Z, Shen J (2016) Charts for estimating rock mass shear strength parameters. Geomech Eng 10(3):257–267

Zheng H, Li T, Shen J, Xu C, Sun H, Lv Q (2018) The effects of blast damage zone thickness on rock slope stability. Eng Geol 246(28):19–27

Chapter 8
A Gas-Mechanical Coupled Hoek-Brown Criterion

Abstract A gas-mechanical coupled constitutive equation is developed for coal containing gas or CO_2 to predict the failure behavior based on fracture mechanics theory. A novel parameter of $\left|\partial \cos\alpha / \partial \sum_1\right|$ has been selected as the failure characteristic parameter for coal containing gas. In addition, a closed-form gas-mechanical coupled constitutive equation is derived. The effects of the coal friction coefficient f, the gas pore pressure p, Biot's coefficient χ and the fracture parameter κ have been coupled in the proposed model. Without considering the effect of pore pressure, the equation wonderfully coincides with the original Hoek-Brown Empirical Strength Criterion. Sensitivity analysis indicates that under a determining confining pressure, the failure stress is an increasing function of f, in contrast, a decreasing function of p, χ and κ. Due to the low tensile strength of coal, the effects of different fracture parameter κ from different fracture criterions can nearly be ignored when the pore pressure p is high. At last, the proposed gas-mechanical coupled model is verified, which is in good agreement with the existing experimental data.

8.1 Introduction

Recently, coal has received renewed attention as a host porous media to absorb CO_2 for greenhouse gas control. It is well known that the change of the physical and mechanical properties of most coal depends on the confining pressure and pore gas pressure, especially for fractured coal. Confining stress and gas pressure have opposite effects on coal properties in many cases. For example, both the failure strength and the elastic modulus of fractured coal increase with the confining stress, but decrease with the pore gas pressure. The permeability and porosity of fractured coal, on the other hand, increase with the pore gas pressure but decrease with the confining stress respectively.

Most investigations have been focused on the limit strength of coal (Walton 1958). Hoek and Brown (1980a, b) proposed an empirical strength criterion for rock and rock masses. The nonlinear criterion includes the uniaxial compressive

J. Zuo and J. Shen, *The Hoek-Brown Failure criterion—From theory to application*,
https://doi.org/10.1007/978-981-15-1769-3_8

strength of the intact rock material and introduces two dimensionless parameters, *m* and *s*.

$$\sigma_1 = \sigma_3 + \sqrt{m\sigma_c\sigma_3 + s\sigma_c^2} \tag{8.1}$$

Hoek-Brown criterion is an empirical formula. *m* and *s* in Hoek-Brown criterion are hardness-softness and fragmentation degree of rock, respectively. Both *m* and *s* are empirical parameters. Kou et al. (1990) considered the effects of numerous microcracks on the mechanical behavior of structural coal and then derived a system of equations to describe the mechanical behavior of structural coal under loading. Unfortunately, few investigations have been focused on the constitutive model for describing the whole deformation and failure process of fractured coal containing gas. Furthermore, most of the previous studies did not consider the effects of pore gas pressure.

From Terzaghi's effective stress theory, the constitutive model should consider the effects of pore gas and fluid pressure on the physical behavior and permeability of porous media. Li (1987) carried out some experiments to discuss the influence of pore gas on the stress-strain curves of coal. Recent continued interests on the single and multiphase flow for fractured coal have been reported in numerous literature (Liu et al. 2011a, b). Recently, carbon dioxide (CO_2) sequestration in deep coal seams has attracted more attention as a method of reducing the output of greenhouse gases to the atmosphere (Saghafi et al. 2007). Therefore, it is very important to develop a rational constitutive model both for coal mining safety and CO_2 sequestration stability.

In the present work, the failure behaviors of fractured coal containing gas have been investigated in detail based on reported experimental data. Then a closed-form constitutive model has been developed to depict the failure criteria for fractured coal that considers the effects of the gas. The proposed failure criterion of fractured coal containing gas generally includes three independent parameters which may be obtained by matching the experimental stress-strain curves. Results obtained with the closed-form constitutive equations based on the fracture mechanics theory will be compared with observed data of a few fractured coal specimens under the coupled effects of different confining pressure and pore gas pressure.

8.2 Typical Stress-Strain Relationship for Fractured Coal

The failure of coal specimens subjected to uniaxial or triaxial compressive stress has been studied in literature (Walton 1958; Zuo et al. 2011). A typical stress-strain curve of coal without gas, collected from the Kailuan Coal Ming Corporation, China, is shown in Fig. 8.1 (Zuo et al. 2011). In this paper, the compressive stress and strain are defined as positive.

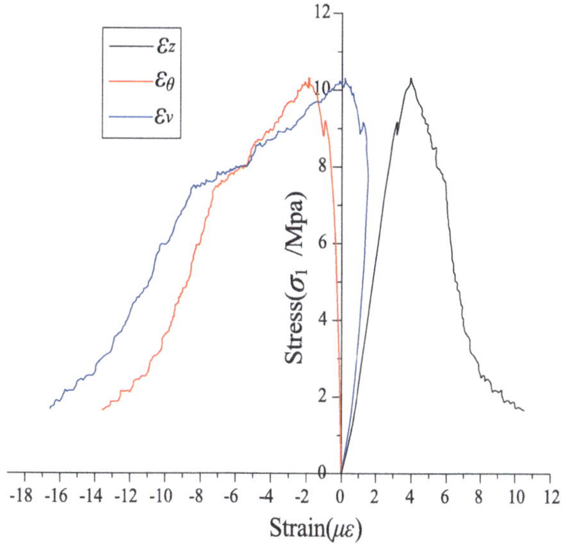

Fig. 8.1 The typical stress-strain curve of coal sample under uniaxial compression stress

Rock brittle fracture under uniaxial compression can usually be viewed as a singular event, occurring after an essentially unrelated elastic deformation, and is characterized uniquely by the peak stress. Some literature also name the peak stress as the ultimate strength or fracture stress (Paterson and Wong 2005). However, since there are a large number of microcracks and cleats distributing stochastically in coal sample, the whole failure process of coal has to be considered as a culmination of a progressive development of the critical microcracks and cleats. Figure 6.1 clearly indicates that the stress-strain curve is approximately linear when the stress is below σ_p which is the proportion limit of coal. Therefore, the initial portion of the stress-strain curve is directly proportional to the strain ε which is known as Hooke's Law ($\sigma = E\varepsilon$). E is the modulus of elasticity, defined by the slope of the straight line portion of the stress-strain curve. With increasing load, the deformation becomes nonlinear elastic when the load reaches σ_e which is the elastic limit of the coal. Though the curve between σ_e and σ_p is nonlinear, the deformation is still recoverable once the external load is removed. This type of deformation involves stretching of the bonds between mineral particles or noncrystalline composition, but no slipping occurs between atoms. In this stage, there is no visible evidence of microcracks in coal. However when the load exceeds σ_e, many microcracks gradually appear. From Fig. 8.1, it can be seen that the stress-strain curve above σ_e is fluctuant which clearly indicates that numerous microcracks are induced by loading.

The peak stress of the coal can also be referred to as the ultimate strength. It is well known that the ultimate strength of coal is enhanced with confining pressure. Brace and Byerlee (1967) showed that dilatant microcracks were strongly oriented parallel to the axis of maximum compression with increased confining pressure. Figure 8.2 is a typical failure mode of coal. Localization of microcracks will result

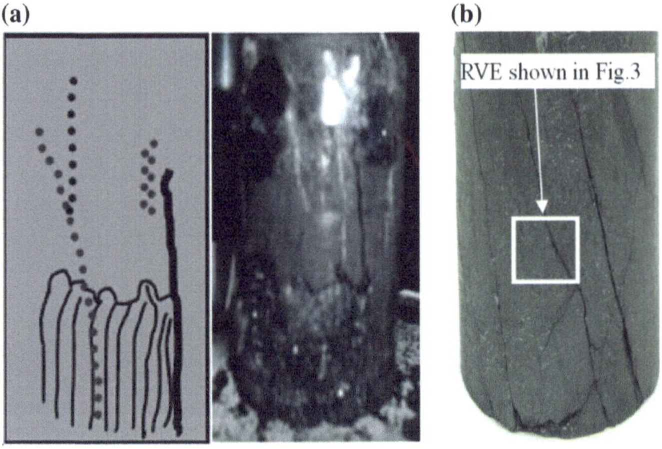

Fig. 8.2 Typical failure mode of coal sample. **a** Indicates that axial splitting is the main failure mechanism of coal under uniaxial compression. **b** Shows that shear failure is the main failure mechanism of coal under triaxial compression (Zuo et al. 2017)

in different macroscopic failure planes under various confining pressures. By comparing the two figures, it can be seen that the distribution zone of microcracks is much larger in the sample under uniaxial compression than that under triaxial compression.

Figure 8.2a indicates that axial splitting is the main failure mechanism of coal under uniaxial compression. Around the bottom of the coal sample, macrocracks (which gradually accumulate with an increasing load and form a large failure zone) are distributed in a direction approximately parallel to the maximum compressive direction. Figure 8.2b shows that shear failure is the main failure mechanism of coal under triaxial compression. Generally speaking, the inclined angle of the shear failure plane relative to the loading axis decreases with increasing confining pressure. In addition, numerous microcracks of various sizes are distributed stochastically in coal samples. Smaller microcracks are greater in number and closer in spacing. Some fundamental conclusions can be obtained from the experimental results and some literature (Griffith 1921; Kachanov 1982; Cotterell and Rice 1980). Under axial compression, larger cracks propagate first. Since there is no confining pressure for uniaxial compression, the microcracks continue to propagate, leading to axial splitting. The increase in confining pressure prevents the propagation of micro-cracks. At a certain stress state, suitably oriented micro-cracks converge in a narrow failure zone. The narrow failure zone eventually becomes a macroscopic failure plane, and leads to a catastrophic fracture. It is obvious that the confining pressure gradually reduces the distribution zone of microcracks in coal. When the microcrack distribution zone reaches a critical value, the coal sample fails catastrophically. Therefore, the failure criterion for coal can be derived based on the distribution zone of microcracks.

8.3 A Model for Fractured Coal Including Gas Based on Hoek-Brown Criterion

Coal contains a large number of microcracks or cleats. Griffith (1921) pointed out that brittle materials are prone to fracturing due to the presence of microcracks. Modern advanced Scanning Electron Microscopic (SEM) and X-ray Computed Tomography (CT) can be conveniently employed to obtain surface and internal microstructure of materials, respectively. The existence of a large number of microcracks has been proven in many brittle materials, including rock and coal (Hemant et al. 2011). These microcracks are usually referred to as fracture or cleat in coal. Since a large amount of coal gas or methane is produced during the coal formation process, the physical and mechanical behaviors of coal are usually subjected to the coupled effects of stress and gas pressure.

Recently, because of continued growth in coal/coal seam gas demands and geological sequestration of CO_2 into coal seams (Cui et al. 2007; Larsen 2004), a number of models have been developed to describe the physical and mechanical behaviors of coal containing gas. Liu and Rutqvist (2010) assumed that coal matrix blocks are connected with each other by the coal-matrix bridges and developed a new coal-permeability model to explicitly consider the fracture-matrix interaction during the coal deformation processes based on the concept of internal swelling stress. Izadi et al. (2011) proposed a mechanistic representation of coal as a collection of unconnected cracks in an elastic swelling medium. The cracks are isolated from each other but swelling within a homogeneous continuum results in a reduction in crack aperture, and a concomitant reduction in permeability. Ma et al. (2011) developed a model based on the volumetric balance between the bulk coal, solid grains and pores, and using the constant volume theory (Massarotto et al. 2009).

As reviewed above, most of these models in the existing literature were based on the Terzaghi's effective stress principle, poroelasticity mechanics, or an equivalent continuum mechanics method. Terzaghi's effective stress principle is viewed as a great success in soil mechanics. However, the cleat or fracture in coal is different with the pores in the soil. Direct application of the effective stress principle to coal containing gas may be very difficult. Poro-elasticity or continuum mechanics methods for coal are also based on the assumption of isotropic, homogeneous, linear stress-strain relationship and infinitesimal deformation in the coal matrix. However, due to the existence of microcracks or cleats, stresses are highly concentrated around microcracks. In this case, fracture mechanics may be more applicable. In the following, a closed-form constitutive equation will be derived for fractured coal containing gas based on fracture mechanics.

The heterogeneity of coal gives rise to various types of tissues, fissures or local stress concentrations that might act as sources for microcracking. As the most likely sources are the existing microcracks, the "sliding crack" has been widely applied to analyze the physical and mechanical behavior of geomaterials (Kachanov 1982; Cotterell and Rice 1980; Zuo et al. 2008). In this paper, a typical sliding crack in an arbitrary plane of unit thickness with the initial length 2a and oriented at an angle β to the direction of the major principal stress σ_1 is adopted, as shown in Fig. 8.3, corresponding to the square zone in Fig. 8.2.

Given six stress components $\sigma_x, \sigma_y, \sigma_z, \tau_{xy}, \tau_{yz}, \tau_{zx}$ applied on the sample, three principal stresses $\sigma_1, \sigma_2, \sigma_3$ for a Representative Volume Element (RVE) can be obtained. The following equation can be solved:

$$\begin{vmatrix} \sigma_x - \sigma & \tau_{xy} & \tau_{xz} \\ \tau_{yx} & \sigma_y - \sigma & \tau_{yz} \\ \tau_{zx} & \tau_{zy} & \sigma_z - \sigma \end{vmatrix} = 0 \tag{8.2}$$

The three eigenvalues of Eq. (8.2) are the three principal stresses $\sigma_1, \sigma_2, \sigma_3$, respectively. For clarification, only the three principal stresses are applied to the RVE of coal sample. It is assumed that the average gas pressure around cleat is p.

Only the external stress, the normal stress σ_β and shear stress τ_β along the cleat are considered, which are given respectively by:

$$\sigma_\beta = \frac{\sigma_1 + \sigma_3}{2} - \frac{\sigma_1 - \sigma_3}{2}\cos 2\beta \tag{8.3a}$$

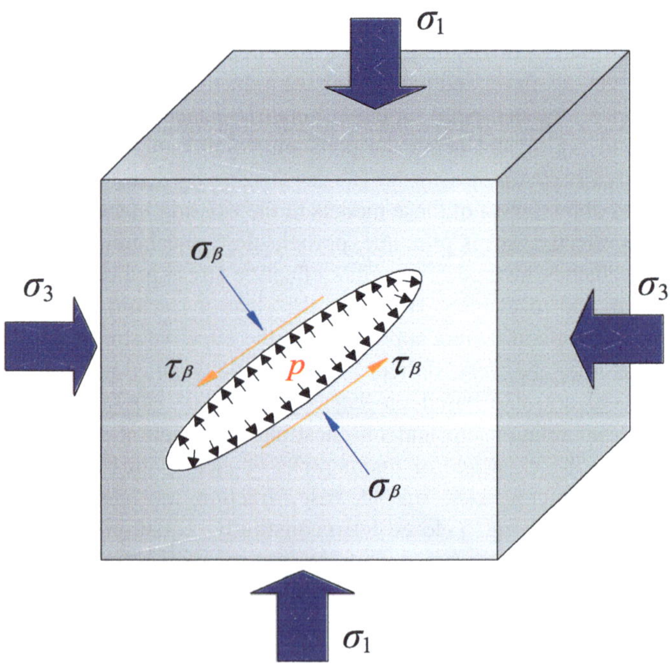

Fig. 8.3 A representative volume element (REV) including a cleat in coal sample (Zuo et al. 2016)

$$\tau_\beta = \frac{1}{2}(\sigma_1 - \sigma_3)\sin 2\beta \tag{8.3b}$$

Since the pore gas pressure is p in the cleat, then there are:

$$\sigma_n = \sigma_\beta - p = \frac{\sigma_1 + \sigma_3}{2} - \frac{\sigma_1 - \sigma_3}{2}\cos 2\beta - p \tag{8.4a}$$

$$\tau_n = \tau_\beta = \frac{1}{2}(\sigma_1 - \sigma_3)\sin 2\beta \tag{8.4b}$$

Since gas cannot sustain shear stress, the factor of pore pressure p is not considered in Eq. (8.4b). Gas may transfer to other phases with increasing pore pressure, such as absorption in the coal seam. In this case, a coefficient χ can be added. Equation (8.4a) can be rewritten.

With the increase of external stress, the stress state around the cleat changes. At a critical stress state, the cleat will deform so as to slide. In reference to the two opposite planes of the cleat, the effective shear stress dictates deformation and movement of the cracked surfaces. The shear stress τ_n arising from the external stress promotes cleat sliding. However, the frictional shear stress on cleat plane arising from the normal stress σ_n prevents the cleat from sliding. Therefore, the effective shear stress can be derived from the above two factors,

$$\tau_e = \tau_n - f\sigma_n = \frac{1}{2}(\sigma_1 - \sigma_3)\sin 2\beta - f\left[\frac{\sigma_1 + \sigma_3}{2} - \frac{\sigma_1 - \sigma_3}{2}\cos 2\beta - \chi p\right] \tag{8.5}$$

where f is the friction coefficient of the cleat plane. If the shear stress τ_n is more than the frictional resistance along the cleat plane, sliding is likely to occur.

Previous studies have indicated the existence of high tensile stress near the tip of any cracks regardless of the sign and magnitude of the external stresses in brittle materials, which has been successfully applied the fracture of glass and rock. In this paper, this concept is also applied to microcracks in coal. Prior to the onset of type I crack propagation, the stress intensity factor K_I at the tip of the crack is (Tada et al. 1973): $K_I = \sigma_t\sqrt{\frac{1}{2}\pi l}$, where σ_t is the tensile strength and l is the crack length. The critical value of K_I can be regarded as the fracture toughness K_{IC}.

In addition, the cleat is also subjected to the effective shear stress τ_e. The crack also belongs to type II crack. Prior to the onset of crack propagation, the stress intensity factor at the initial crack tip is $K_{II} = \tau_e\sqrt{\frac{1}{2}\pi l}$.

During microcrack propagation in geomaterials, the fracturing path is generally curved due to non-uniform stress fields. External loads may induce wing cracks at the tips of sliding microcracks based on a great number of experimental investigations and theories (Brace and Bombolakis 1963; Kachanove 1982; Cotterell and Rice 1980). If the kink length l' is infinitesimal, the stress intensity factor at the tip of the secondary crack based on linear elastic fracture mechanics can be expressed by (Kachanove 1982; Zuo et al. 2008)

$$\mathrm{K_{II}} \geq \kappa \mathrm{K_I} \tag{8.6}$$

κ is a constant which can be derived from the various approximations suggested in the literature on the kinked crack. It is different corresponding to different fracture criterion. For example, in the maximum hoop stress criterion proposed by Erdogan and Sih (1963), κ is approximately equal to $\frac{2}{\sqrt{3}}$. In the minimum strain energy criterion by Sih (1974) and Sih and Macdonald (1974), κ is approximately equal to $\sqrt{\frac{2+2v-v^2}{[3(1-v)]}}$ where v is Poisson's ratio. In the maximum energy release rate criterion by Nuismer (1975), κ is approximately equal to 1. If the microcrack propagates subjected to an external stress, substitution of $\mathrm{K_{II}} = \tau_e\sqrt{\frac{1}{2}\pi l}$ and $\mathrm{K_I} = \sigma_t\sqrt{\frac{1}{2}\pi l}$ in Eq. (8.6) leads to:

$$\tau_e \geq \kappa \sigma_t \tag{8.7}$$

In order to identify the critical orientation angle β for the sliding cleat, the first-order and second-order derivations of $\mathrm{K_{II}}$ corresponding to β, are calculated, respectively,

$$\frac{\partial K_{II}}{\partial \beta} = \sqrt{\frac{1}{2}\pi l}\frac{\partial \tau_e}{\partial \beta} = 0 \tag{8.8a}$$

$$\frac{\partial^2 K_{II}}{\partial \beta^2} < 0 \tag{8.8b}$$

The second-order derivation or Eq. (8.8b) shall be less than 0, corresponding to the maximum value of $\mathrm{K_{II}}$. As the pore pressure is independent of the orientation angle β, the critical orientation angle β_0 can be derived, using Eqs. (8.5) and (8.8a, 8.8b)

$$\beta_0 = \frac{1}{2}\arctan(1/f) \tag{8.9}$$

8.4 Microscopic Observations on the Crack Distribution of Coal

The micro/mesoscale studies on the brittle failure process of coal have developed rapidly in recent years, in terms of advanced observation techniques such as optical microscopic study, scanning electron microscope(SEM), and stereological techniques (for example, X-ray CT).

Figure 8.4 indicates that the failure modes are different for coal specimens under different confining pressure. Under uniaxial compression, the main failure mechanism of coal is the cataclastic mechanism. For weakly cohesive coal, cataclasis arises mainly from coal grain crushing and catastrophic failure, as modeled with Hertzian fracture mechanism (Zhang et al. 1990; Wong et al. 2004) and the catastrophe theory, respectively. The growth of intragranular cracks in the porous coal may either dilate or compact the pore space. Intragranular cracks that initiate from grain contacts to propagation across an initially intact grain can break the coal specimen into many fragments, as shown in Fig. 8.4a. With the increase of confining pressure, the main failure mechanism will change. When the confining pressure reaches 5 MPa, many parallel microcracks can be observed in the cross-section, as shown in Fig. 8.4b. The suppression of longitudinal splitting occurs at the relatively low confining pressure of 5 MPa. When the confining pressure increases to 20 MPa, the parallel microcracks coalesce to be a macroscopic shear failure surface in coal, as shown in Fig. 8.4c. Shear failure is the main failure mechanism in this case. Therefore, the transition from brittle cataclasis to shear failure with increasing confining pressure from 0 to 20 MPa has been observed in experiments on coal samples. Such a phenomenon is also supported by experimental observations in many rocks. Under higher confining pressure, the brittle-ductile transition occurs in coal. For different rocks, the pressure corresponding to the transition is different, such as 20–30 MPa for pure coarse-grained marbles, 100 MPa for fine-grained Solnhofen limestone (Heard 1960). Vajdova et al. suggested that the transition pressure tends to decrease with increasing porosity for limestone (Vajdova et al. 2004).

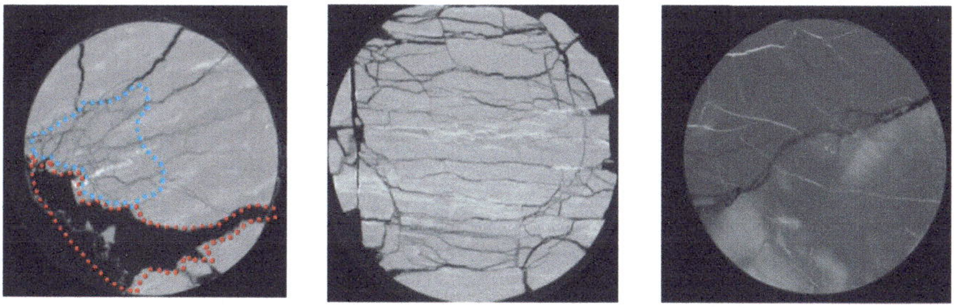

(a) Uniaxial compression (b) Confining pressure 5MPa (c) Confining pressure 20MPa

Fig. 8.4 Typical CT image of coal under different confining pressure

Figure 8.4a shows that intragranular cracks that initiate from grain contacts to the propagation across an initially intact grain can break the coal specimen into many fragments. Figure 8.4b shows that when the confining pressure reaches 5 MPa, many parallel microcracks can be observed in the cross section. Figure 8.4c shows that when the confining pressure increases to 20 MPa, the parallel microcracks coalesce to be a macroscopic shear failure surface in coal.

The above analysis indicates that shear localization develops by coalescence of microcracks. Therefore, it is assumed that all the randomly distributed microcracks in coal, which satisfy Eq. (8.7), are converged to a point O. A fan-shaped area of crack distribution zone with a radius of $2a$ can be obtained, as shown Fig. 8.5. The included angle α of the fan section is defined as the "micro-failure orientation angle" which is occupied by all of the fitting microcracks, as shown in Fig. 8.5.

Based on the tests on coal, α is a monotonic decreasing function of confining pressure. For simplicity, $\sum_1$ and $\sum_3$ are used to represent σ_1/σ_t and σ_3/σ_t, respectively. Substituting Eq. (8.5) into Eq. (8.7), we have:

$$\left(f\sum_1 + \kappa + \frac{f\chi p}{\sigma_t}\right)\tan^2\beta - \left(\sum_1 - \sum_3\right)\tan\beta + \left(f\sum_3 + \kappa + \frac{f\chi p}{\sigma_t}\right) \le 0 \quad (8.10)$$

The tangent of β can then be determined by

$$\tan\beta_1 \le \tan\beta \le \tan\beta_2 \quad (8.11)$$

where

$$\tan\beta_1 = \frac{(\sum_1 - \sum_3) - \sqrt{(\sum_1 - \sum_3)^2 - 4(f\sum_1 + \kappa + \frac{f\chi p}{\sigma_t})(f\sum_3 + \kappa + \frac{f\chi p}{\sigma_t})}}{2(f\sum_1 + \kappa + \frac{f\chi p}{\sigma_t})} \quad (8.12)$$

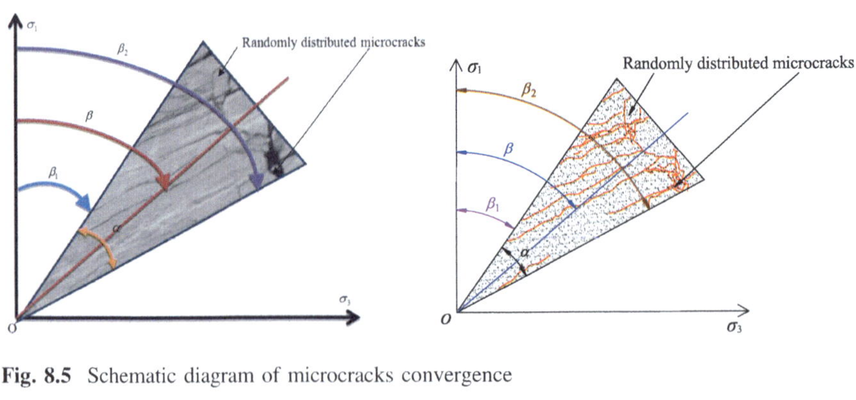

Fig. 8.5 Schematic diagram of microcracks convergence

$$\tan\beta_2 = \frac{(\sum_1 - \sum_3) + \sqrt{(\sum_1 - \sum_3)^2 - 4(f\sum_1 + \kappa + \frac{f\chi p}{\sigma_t})(f\sum_3 + \kappa + \frac{f\chi p}{\sigma_t})}}{2(f\sum_1 + \kappa + \frac{f\chi p}{\sigma_t})} \tag{8.13}$$

Using Eqs. (8.12) and (8.13), one can obtain

$$\tan(\beta_1 + \beta_2) = \frac{1}{f} = \tan 2\beta_0 \tag{8.14}$$

i.e.

$$\beta_1 + \beta_2 = 2\beta_0 \tag{8.15}$$

Transformation of the equation form leads to:

$$\beta_0 - \beta_1 = \beta_2 - \beta_0 \tag{8.16}$$

Equation (8.16) indicates that the propagation angle domain of microcracks is symmetrical in the critical crack orientation. Since the fan-shaped included angle α is equal to $\beta_2 - \beta_1$, we have

$$\tan\alpha = \frac{\sqrt{(\sum_1 - \sum_3)^2 - 4(f\sum_1 + \kappa + \frac{f\chi p}{\sigma_t})(f\sum_3 + \kappa + \frac{f\chi p}{\sigma_t})}}{f(\sum_1 + \sum_3) + 2(\kappa + \frac{f\chi p}{\sigma_t})} \tag{8.17}$$

8.5 Constitutive Equation for Coal Containing Pressure Gas

Though coal failure is very complicated under external load, the characteristic parameter should be an invariant corresponding to macroscopic failure. In recent 50 years, many failure criteria have been proposed for rock materials. The weakest link theories have been applied for brittle failure in cases of uniaxial tension (Brown 1971) and compressive conditions (Lundborg 1972). However, it is a statistical theory. Brady (1969) proposed a statistical theory based on the proliferation of microcracking, assuming that failure occurs "when the volumetric strain due to microcracking attains a critical value" (Wiebols and Cook 1968).

However, many experiments have indicated that initiation of an internal microcrack does not mean rock failure (Paterson and Wong 2005). Localized coalescence of microcracks that extends over two or more mineral grains is quite evident in granite (Wong 1982; Moore and Lockner 1995) and gabbro (Wong and Biegel 1985) subjected to stresses beyond the peak stress. The coalescence of many microcracks induces the final failure of rock. When microcracks grow in non-uniform stress fields, the path of the fracture is generally curved based on fracture mechanics (Cotterell and Rice 1980; Kachanov 1982). Much study efforts have been applied to crack initiation at an angle to a pre-existing crack. The confining pressure σ_3 can effectively restrain wing microcracks from propagating arbitrarily. The wing crack propagates optimally in the direction parallel to the orientation of major principal stress σ_1 with increasing σ_3. Therefore, crack initiation cannot represent rock failure. Therefore, it is not suitable to establish a failure criterion just from microcrack initiation.

When the density of internal microcracks reaches a critical value, fracturing occurs. However, the experimental results indicate that the density of micro-cracks is not a material constant since it is affected by confining pressure. It is also related to coal types. An increase in confining pressure σ_3 leads to a decrease in the density of microcracks. Therefore, coal failure characteristic parameter should be relevant to the confining pressure σ_3. Also, it is related to the orientation angle α of microcracks. Experimental results also indicate that the higher the confining pressure σ_3 (namely, the dimensionless parameter $\sum_3$), the smaller the microcrack orientation included angle α.

It is assumed that any dimensionless parameter of $\overset{i}{\sum}_3$ (i=0, 1, 2, ..., n) at the moment of coal failure corresponds to a microcrack orientation included angle $\overset{i}{\alpha}(i = 0, 1, 2, \ldots, n)$ and the dimensionless parameter $\overset{0}{\sum}_3 = 0$. Obviously, the micro-failure orientation included angle α is complicated and will change with varying confining pressure σ_3. Therefore, α cannot be regarded as the failure characteristic parameter directly. Figure 8.6 schematically indicates that how α changes with the confining pressure σ_3. The left image in Fig. 8.6 indicates that α decreases with increasing σ_3, but increases with increasing σ_1. The right image in Fig. 8.6 indicates that, at a critical stress state with different σ_1 and σ_3, the same crack distribution angle α can be obtained. Cosine calculation of α indicates the development trend of the propagation direction of crack with the increasing of confining pressure σ_3.

Based on Eq. (8.17), the cosine of the micro-failure orientation angle α can be obtained:

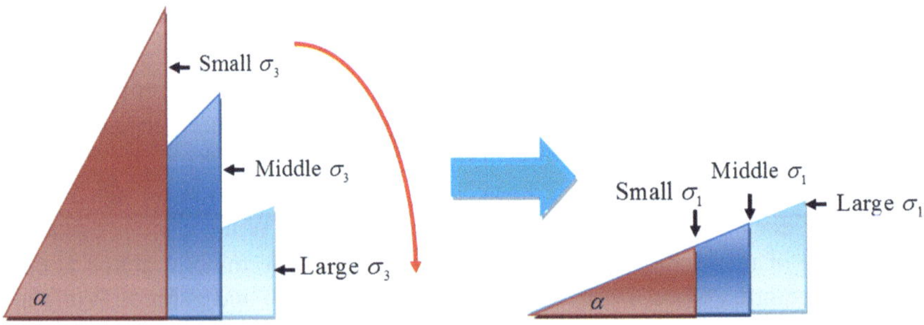

Fig. 8.6 Schematic for deformation under different confining pressure and axial stress

$$\cos\alpha = \frac{f}{\sqrt{1+f^2}}\left(1 + 2\frac{\sum_3 + (\kappa + \frac{f\chi p}{\sigma_t})/f}{\sum_1 - \sum_3}\right) \tag{8.18}$$

For a certain confining pressure, coal fractures at a specific failure angle at a definite axial compression stress. Therefore, there is a critical relationship between $\cos\alpha$ and the dimensionless parameter $\sum_1$. It is well known that coal fractures with increasing σ_1. The propagation and coalescence of the initial microcracks (just a few or several) induce a finial macroscopic crack (which is composed of a large number of microcracks). Therefore, $\cos\alpha$ is the monotonic decreasing function of the dimensionless parameter $\sum_1$. In contrast, α is constrained by the confining pressure σ_3. In other words, $\cos\alpha$ is the monotonic increasing function of the dimensionless parameter $\sum_3$.

Therefore, it is anticipated that the micro-failure orientation angle α decrease with increasing confining pressure σ_3. It also indicates that $\cos\alpha$ increase with $\sum_3$ and coal gas pressure p. Differentiating Eq. (8.18) with respect to $\sum_1$, one can obtain:

$$\left|\frac{\partial\cos\alpha}{\partial\sum_1}\right| = \frac{f}{\sqrt{1+f^2}}\frac{2\sum_3 + 2(\kappa + \frac{f\chi p}{\sigma_t})/f}{(\sum_1 - \sum_3)^2} \tag{8.19}$$

where $|\partial\cos\alpha/\partial\sum_1|$ is defined as the rate of change of $\cos\alpha$ to the axial stress dimensionless parameter $\sum_1$.

Combining Eqs. (8.18) and (8.19), we have:

$$\left|\frac{\partial\cos\alpha}{\partial\sum_1}\right| = \frac{f}{\sqrt{1+f^2}}\frac{[(\sqrt{1+f^2}\cos\alpha)/f - 1]^2}{2\sum_3 + 2(\kappa + \frac{f\chi p}{\sigma_t})/f} \tag{8.20}$$

We think that the failure characteristic parameter $|\partial\cos\alpha/\partial\sum_1|$ is always a constant when the rock fractures, no matter what the value of $\sum_3$ is. So $|\partial\cos\alpha/\partial\sum_1|$ can be determined by simple uniaxial compression tests. For uniaxial compression tests, $\sum_3 = 0$, and $\sum_1 = \sigma_c/\sigma_t = \sum_c$. The failure characteristic parameter $|\partial\cos\alpha/\partial\sum_1|$ is:

$$\left|\frac{\partial\cos\alpha}{\partial\sum_1}\right| = \frac{f}{\sqrt{1+f^2}}\frac{2(\kappa + \frac{f\chi p}{\sigma_t})}{f\sum_c^2} \tag{8.21}$$

For any known $f, \kappa, \chi, \sigma_t, \sigma_c$ and gas pressure p of coal, there is a characteristic constant $|\partial\cos\alpha/\partial\sum_1|$. From Eqs. (8.18) and (8.20), the following equation is derived:

$$\sum_1 = \sum_3 + \sqrt{\frac{f}{\kappa + \frac{f\chi p}{\sigma_t}} \sum_c^2 \sum_3 + \sum_c^2} \tag{8.22}$$

Substituting the dimensionless parameters $\sum_1 = \sigma_1/\sigma_t$ and $\sum_3 = \sigma_3/\sigma_t$ into Eq. (8.22), then

$$\sigma_1 = \sigma_3 + \sqrt{\frac{f}{\kappa + \frac{f\chi p}{\sigma_t}} \frac{\sigma_c}{\sigma_t} \sigma_c \sigma_3 + \sigma_c^2} \tag{8.23}$$

A close-formed constitutive equation for coal containing gas is derived theoretically from fracture mechanics based on micro-failure phenomena. It is wonderfully similar to the original Hoek-Brown empirical criterion (Hoek and Brown 1980a, b). Equation (8.23) clearly indicates that the failure stress σ_1 is a monotonic increasing function of confining pressure σ_3, but a monotonic decreasing function of pore pressure p of coal.

It also indicates that the failure stress σ_1 is related to the friction coefficient f, the coefficient κ of mixed-mode fracture criterion, uniaxial compressive strength σ_c and uniaxial tensile strength σ_t of coal sample. This relationship is also consistent with experimental data from coal containing gas. It is also consistent with Terzzaghi's theory.

The above analysis shows that $|\partial \cos \alpha / \partial \sum_1|$ can be regarded as the failure characteristic parameter for coal. The other trigonometric functions of α with respect to the dimensionless parameter $\sum_1$ (such as $|\partial \sin \alpha / \partial \sum_1|$, $|\partial \tan \alpha / \partial \sum_1|$, $|\partial \csc \alpha / \partial \sum_1|$ and $|\partial \sec \alpha / \partial \sum_1|$) cannot be chosen as the failure characteristic parameter because they cannot remain constant with a change of the confining pressure σ_3 when coal fractures.

8.6 Sensitive Analyses and Comparison with Experimental Data

A close-formed constitutive relationship for the failure of coal including gas has been derived in Sect. 8.5. In this section, the effect of parameters in Eq. (8.23) will be analyzed, and the relationships are validated by comparing the theoretical results with selected laboratory data. For coal sample used in the experiment, $\sigma_c = 4.54\text{MPa}$, $\sigma_t = 0.1\text{MPa}$. According to different failure criterion, there are different κ (Erdogan and Sih 1963; Sih 1974; Sih and Macdonald 1974; Nuismer 1975), such as 1.155, 1 and $\sqrt{(2 + 2v - v^2)/[3(1 - v)]}$. If Poisson's ratio of coal ranges from 0.1 to 0.5, then κ is in the range of about 0.9–1.35. Therefore, the effects of κ from 0.1 to 2 on the failure stress are analyzed. Figure 8.7 shows the schematic figures when $f = 0.5$, $\chi = 1$ and gas pressure p from 0 to 4 MPa. It clearly indicates

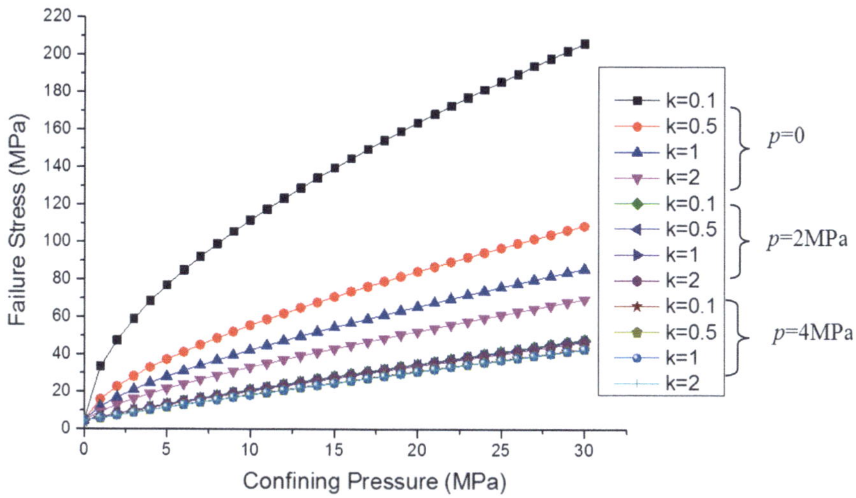

Fig. 8.7 The effects of pore pressure p and fracture coefficient κ on the failure criterion of coal

that the fracture coefficient κ has great effects on failure stress when p = 0. The bigger κ is, the smaller the failure stress is. However, when there is pore pressure, for example p = 2 or 4 MPa, the effect of κ greatly decreases. With the increase of p, the failure stress changes nearly from nonlinear to linear with the confining pressure. Therefore, the failure criterion must consider the effect of κ from different fracture criterion when there is no pore pressure in coal. In contrast, when there is pore pressure in coal, the failure criterion need not consider the effect of κ. The main reason could be that the effect of p is more significant than that of κ.

Figure 8.8 shows the effects of pore pressure p and friction coefficient f on the failure criterion of coal when $\kappa = 1$ and $\chi = 1$. It clearly indicates that the friction coefficient f has also great effects on failure stress when $p = 0$. The bigger f is, the higher the failure stress is. However, when there is pore pressure, for example $p = 2$ or 4 MPa, the effect of f has greatly decreased. With the increase of p, the failure stress changes nearly from nonlinear to linear with the confining pressure. Therefore the failure criterion must consider the effect of f from different fracture criterion when there is no pore pressure in coal. In contrast, when there is pore pressure in coal, the failure criterion need not consider the effect of f when p is bigger than 4 MPa. The main reason could be that the effect of p is more significant than that of f.

Figure 8.9 shows the effects of pore pressure p and coefficient χ on the failure criterion of coal when $\kappa = 1$ and $f = 0.5$. It clearly indicates that the coefficient χ has also great effects on failure stress when there is pore pressure. The bigger χ is, the smaller the failure stress is. However, when p = 0, the effect of χ has no effect. Therefore the failure criterion must consider the effect of χ from different fracture criterion when there is pore pressure in coal. In contrast, when there is no pore pressure in coal, the failure criterion need not consider the effect of χ.

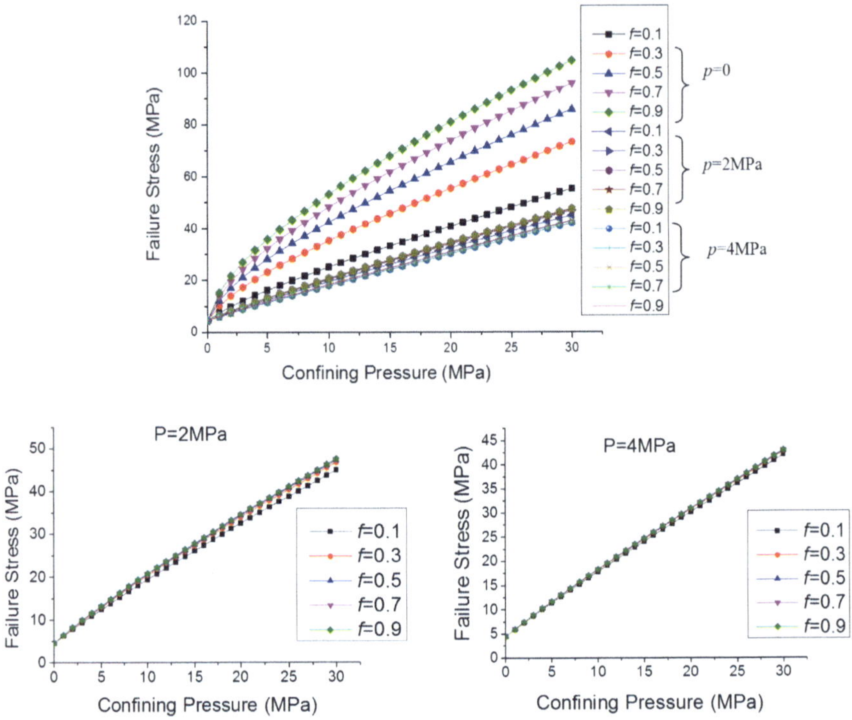

(a) enlarged figure(p=2MPa) (b) enlarged figure(p=4MPa)

Fig. 8.8 The effects of pore pressure p and friction coefficient f on the failure criterion of coal when $\kappa = 1$ and $\chi = 1$

In this section, only the effect of the coefficient $\lambda = f/(\kappa + (f\chi p/\sigma_t))$ is discussed. If λ is changed from 0.01 to 2, Fig. 8.10 can be obtained. It clearly indicates that with increasing confining pressure σ_3, the finial failure stress σ_1 increases nonlinearly. The whole coefficient λ also changes the variability ratio of failure stress. In addition, the larger the coefficient λ, the faster the failure stress σ_1 increases (Fig. 8.10).

A set of test results (Yao and Zhou 1988) are used to verify the failure criterion (Eq. 8.23). The average axial compression strength and tensile strength of the original coals are 14 MPa and 0.2 MPa, respectively. Barton and Choubey (1977), Byerlee (1978), Einstein and Dowding (1981), and Lockner (1995) reported that the range of rock friction coefficient is from about 0.2 to 0.8, and the majority falling within 0.4–0.7. Considering the property of the coal, the coal friction coefficient is assumed to be about 0.7. Based on the maximum hoop stress criterion proposed by Erdogan and Sih (1963), $\kappa = 2/\sqrt{3}$. Assume Biot's coefficient χ to be 0.1 for coal.

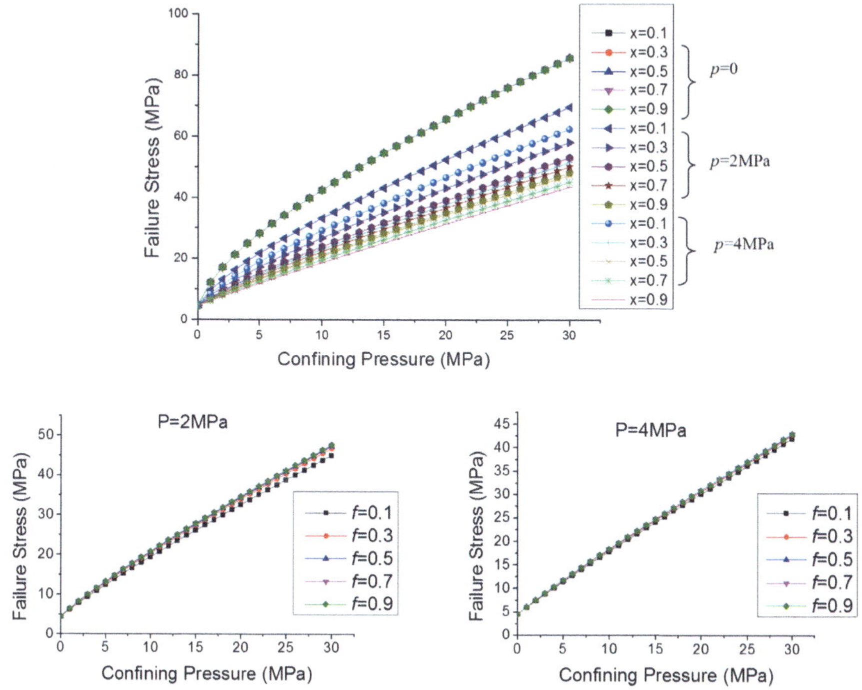

(a) enlarged figure(p=2MPa) (b) enlarged figure(p=4MPa)

Fig. 8.9 The effects of pore pressure p and the coefficient χ on the failure criterion of coal

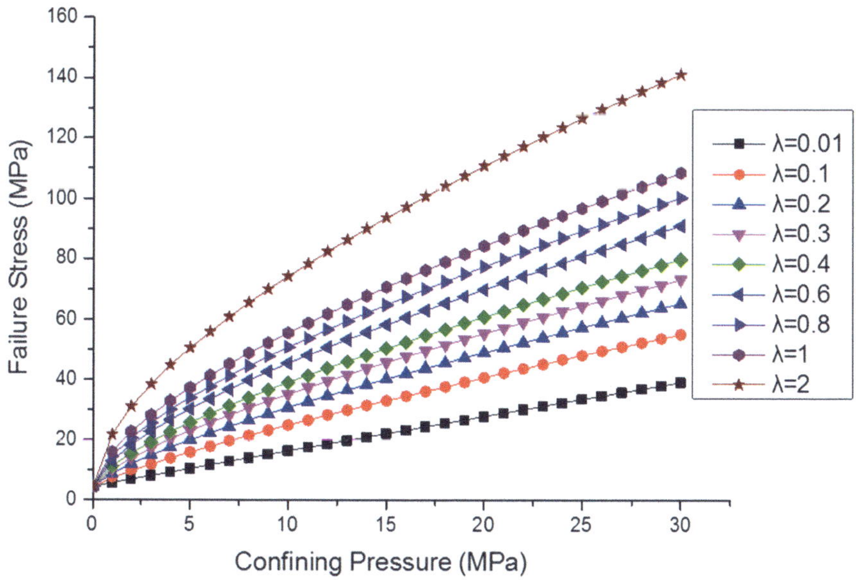

Fig. 8.10 The effects of $\lambda = f/(\kappa + (f\chi p/\sigma_t))$ on the failure criterion of coal

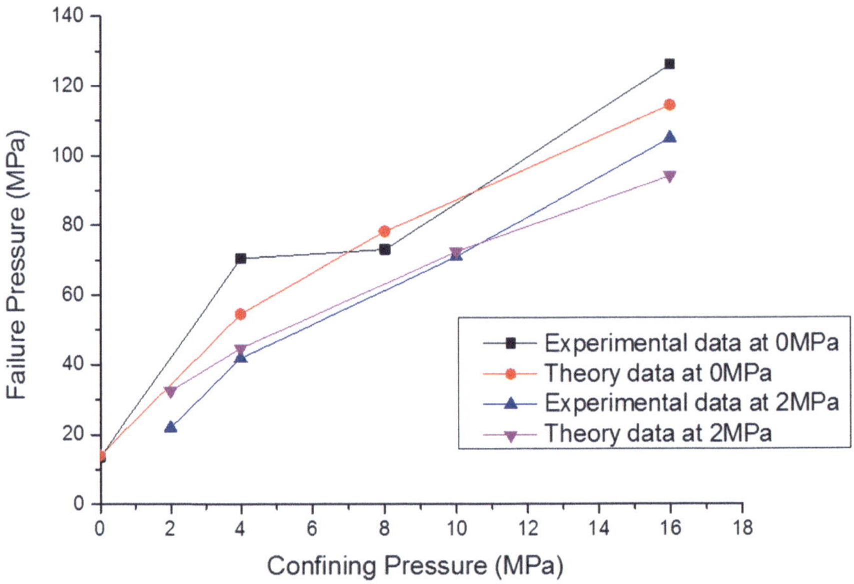

Fig. 8.11 Comparison between experimental data with calculation results by Eq. (8.23)

Comparison between experimental data and the calculation results by Eq. 8.23 is shown in Fig. 8.11. It is clearly shown that Eq. 8.23 fits very well with experimental data.

References

Barton NR, Choubey V (1977) The shear strength of rock joints in theory and practice. Rock Mech Rock Eng 10(1–2):1–54

Brace WF, Bombolakis EG (1963) A note on brittle crack growth in compression. J Geophys Res 68:3709–3713

Brace WF, Byerlee JD (1967) Recent experimental studies of brittle fracture of rocks. In: Fairhurst C (ed) Proceedings 8th symposium on rock mechanics. University Minnesota, Sept. 1966. A.I.M.E., New York, pp 58–81

Brady BT (1969) A statistical theory of brittle fracture for rock materials. Part I. Brittle failure under homogeneous axisymmetric states of stress. Int J Rock Mech Min 6:21–42

Brown ET (1971) Strength-size effects in rock material. In: Rock fracture. Proceedings of the international symposium on rock mechanics, Nancy, 1971, pp Paper II-11 (11 pp)

Byerlee J (1978) Friction of rocks. Pure Appl Geophys 116(4–5):615–626

Cotterell B, Rice JR (1980) Slightly curved or kinked cracks. Int J Fracture 16(2):155–169

Cui X, Bustin R, Chikatamarla L (2007) Adsorption-induced coal swelling and stress: implications for methane production and acid gas sequestration into coal seams. J Geophys Res 112:B10202

Einstein HH, Dowding CH (1981) Shearing resistance and deformability of rock joints. Phys Prop Rocks Miner pp 177–220

Erdogan F, Sih GC (1963) On the crack extension in plates under plane loading and transverse shear. J Basic Eng 85:519–527

Griffith AA (1921) The phenomena of rupture and flow in solids. Philos Trans R Soc Lond Ser A 221:163–198

Heard HC (1960) Transition from brittle fracture to ductile flow in Solenhofen limestone as a function of temperature, confining pressure, and interstitial fluid pressure. In: Griggs D, Handin J (eds) Rock deformation. Geol Soc Am Mem 79:193–226

Hemant K, Edward L, Sam K, Richard B et al (2011) Inducing fractures and increasing cleat apertures in a bituminous coal under isotropic stress via application of microwave energy. Int J Coal Geol 88(1):75–82

Hoek E, Brown ET (1980a) Underground excavations in rock. The Institution of Mining and Metallurgy, London

Hoek E, Brown ET (1980b) Empirical strength criterion for rock mass. J Geo-Tech Eng Div 106 (9):1013–1035

Izadi G, Wang S, Elsworth D, Liu J, Wu Y, Pone D (2011) Permeability evolution of fluid-infiltrated coal containing discrete fractures. Int J Coal Geol 85:202–211

Kachanov ML (1982) A microcrack model of rock inelasticity, part II: propagation of microcracks. Mech Mater 1:29–41

Kou S, Ye D, Ding Y (1990) A constitutive model for structural coal based on fracture and damage mechanics. Eng Fract Mech 35(415):835–843

Larsen J (2004) The effects of dissolved CO2 on coal structure and properties. Int J Coal Geol 57:63–70

Li Z (1987) Study on mechanism of coal and gas outburst in in-seam drivage. Chinese J Coal 1:17–27

Liu H, Rutqvist J (2010) A new coal permeability model: internal swelling stress and fracture–matrix interaction. Transp Porous Media 82(1):157–171

Liu J, Wang J, Chen Z, Wang S, Elsworth D, Jiang Y (2011a) Impact of transition from local swelling to macro swelling on the evolution of coal permeability. Int J Coal Geol 88(1):31–40

Liu Jishan, Chen Zhongwei, Elsworth Derek, Hongyan Qu, Chen Dong (2011b) Interactions of multiple processes during CBM extraction: a critical review. Int J Coal Geol 87(3–4):175–189

Lockner DA (1995) Rock failure. In: Rock physics and phase relations: a handbook of physical constants, 3:127–147

Lundborg N (1972) A statistical theory of the polyaxial compressive strength of materials. Int J Rock Mech Min 9:617–624

Massarotto P, Golding S, Rudolph V (2009) Constant volume CBM reservoirs: an important principle. In: International coalbed methane symposium. Tuscaloosa, Alabama

Ma Q, Harpalani S, Liu S (2011) A simplified permeability model for coalbed methane reservoirs based on matchstick strain and constant volume theory. Int J Coal Geol 85(1):43–48

Moore DE, Lockner DA (1995) The role of microcracking in shear-fracture propagation in granite. J Struct Geol 17:95–114

Nuismer RJ (1975) An energy release rate criterion for mixed mode fracture. Int J Fract 11(2):245–250

Paterson MS, Wong TF (2005) Experimental rock deformation-the brittle field. Springer

Saghafi A, Faiz M, Roberts D (2007) CO2 storage and gas diffusivity properties of coals from Sydney Basin, Australia. Int J Coal Geol 70:240–254

Sih GC (1974) Strain-energy-density factor applied to mixed mode crack problems. Int J Fract 10 (3):305–321

Sih GC, Macdonald B (1974) Fracture mechanics applied to engineering problems–strain-energy density fracture criterion. Eng Frac Mech 6(2):361–386

Tada I, Paris PC, Irwin GR (1973) The stress analysis of cracks handbook. Del Res Corp, Hellertown

Vajdova VP, Baud P, Wong T-f (2004) Compaction, dilatancy and failure in porous carbonate rocks. J Geophys Res 109(B5):1–16

Walton WH (1958) Mechanical properties of non-metallic brittle materials. Butterworth, London, pp 5–28, 123–145

Wiebols GA, Cook NGW (1968) An energy criterion for the strength of rock in polyaxial compression. Int J Rock Mech Min 5:529–549

Wong T-f (1982) Micromechanics of faulting in Westerly granite. Int J Rock Mech Min 19:49–64

Wong T-f, Biegel R (1985) Effects of pressure on the micromechanics of faulting in San Marcos gabbro. J Struct Geol 7:737–749

Wong T-f, David C, Menéndez B (2004) Mechanical compaction. In: Guéguen Y, Boutéca M (eds) Mechanics of fluid-saturated rocks. Elsevier Academic Press, Amsterdam, pp 55–114

Yao Y, Zhou S (1988) The mechanical property of coal containing gas. J China Univ Min Technol 1:1–7

Zhang J, Wong T-f, Davis DM (1990) Micromechanics of pressure-induced grain crushing in porous rocks. J Geophys Res 95:341–352

Zuo J, Li H, Xie H, Ju Y, Peng S (2008) A nonlinear strength criterion for rocklike materials based on fracture mechanics. Int J Rock Mech Min Sci 45(4):594–599

Zuo J, Xie H, Wu A, Liu J (2011) Investigation on failure mechanisms and mechanical behavior of deep coal-rock single body and combined body. Chin J Rock Mechan Eng 30(1):84–92

Zuo J, Jiang G, Wei W, Su H (2016) A gas-mechanical coupled constitutive equation for fractured coal containing gas. J Nat Gas Sci & Eng 36:1109–1119

Zuo J, Chen Y, Wang C (2017) Failure mechanics and models for deep coal-rock combined bodies. Science Press, p 47

Chapter 9
An UCS Model for Anisotropic Blocky Rock Masses Satisfying the Hoek-Brown Criterion

Abstract The Hoek–Brown (HB) failure criterion is one of the most widely used failure criteria in rock engineering. Based on the Geological Strength Index (GSI) system, a number of empirical models have been proposed in parallel with this criterion to estimate the strength and deformation properties of rock masses such as uniaxial compressive strength (UCS) and deformation modulus. However, the GSI system does not incorporate the effects of joint orientation on the quality of a rock mass. This means that these empirical models cannot capture anisotropic rock mass strength caused by joint orientations. In this research, UDEC rock mass models, which are calibrated by laboratory data, are used to investigate the effects of joint orientation on rock mass strength in an unconfined state. The values of UCS obtained from the numerical simulation are then compared with those calculated from traditional empirical UCS models based on the GSI system. The comparison study shows that the value of UCS is significantly overestimated by the traditional empirical model when $10° < \beta < 45°$, which will have serious safety implications for engineering designs. To rectify the problem, based on the analysis of numerical simulation results, an anisotropic weighting factor f_β is proposed to be used to refine the empirical UCS model. The modified UCS model is demonstrated to be capable of giving conservative but more accurate prediction of the rock mass strength for various joint orientations, which will result in more optimal and safer engineering designs.

9.1 Introduction

One of the most difficult problems encountered in the design of rock engineering structures is the determination of strength and deformation properties of rock masses containing joints, bedding planes, and other geotechnical defects (Shen et al. 2012; Wang et al. 2017). Currently, one of the widely used empirical relations in rock engineering is the generalized Hoek–Brown (HB) failure criterion (Hoek et al. 2002; Hoek and Brown 2018) that characterizes the stress conditions that lead to failure of intact rock and rock masses, as shown in Eq. (9.1), where σ_1 and σ_3 are

J. Zuo and J. Shen, *The Hoek-Brown Failure criterion—From theory to application*,
https://doi.org/10.1007/978-981-15-1769-3_9

the maximum and minimum principal stresses, respectively, and m_i, σ_{ci}, the Geological Strength Index (GSI) and disturbance factor D are the HB input parameters. The original HB failure criterion, which was proposed by Hoek and Brown (1980) for estimating intact rock strength, requires two intact rock properties, namely, the uniaxial compressive strength (UCS or σ_{ci}) and a constant m_i of the intact rock. It was extended to estimate the rock mass strength using the GSI and D (Hoek et al. 2002).

$$\sigma_1 = \sigma_3 + \sigma_{ci}\left(m_b \frac{\sigma_3}{\sigma_{ci}} + s\right)^a \tag{9.1}$$

where, m_b, s and a are input parameters, and can be calculated from Eqs. (9.2) and (9.4).

$$m_b = m_i \exp\left(\frac{GSI - 100}{28 - 14D}\right) \tag{9.2}$$

$$s = \exp\left(\frac{GSI - 100}{9 - 3D}\right) \tag{9.3}$$

$$a = \frac{1}{2} + \frac{1}{6}\left(e^{-GSI/15} - e^{-20/3}\right) \tag{9.4}$$

The HB failure criterion has been extensively used by many researchers (Jimenez et al. 2008; Shen and Karakus 2014; Zuo et al. 2015; Vásárhelyi et al. 2016; Wang and Shen 2017; Zhang et al. 2018; Vorobiev and Rubin 2018) to estimate the rock mass strength. A number of empirical models have been proposed in parallel with the HB failure criterion to estimate the rock masses properties based on the GSI system (Cai et al. 2004; Kim et al. 2007; Hoek and Brown 2018). It is important to note that the GSI system (Hoek et al. 2013) does not incorporate the effects of joint orientation on the quality of a rock mass. However, laboratory test results (Brown and Trollope 1970; Yang et al. 1998; Kulatilake et al. 2001; Singh et al. 2002; Ghazvinian and Hadei 2012; Kulatilake et al. 2015; Cao et al. 2016; Wang et al. 2018) using physical models with rock-like materials show that the strength and deformation properties corresponding to the failure modes of rock masses with two joint sets are affected significantly by the orientations of the joints with respect to the loading direction. Therefore, empirical models based on the GSI system must be used with great caution for blocky rock masses containing cube blocks created by intersecting discontinuity sets.

The failure mode of a jointed rock mass is complicated and it influences the rock mass properties (Dershowitz and Einstein 1988; Darlington et al. 2011; Alshkane et al. 2017). To have a better understanding of the occurrence of various modes of failure and their effects on the strength of jointed rock masses, Singh et al. (2002) carried out laboratory tests using rock mass specimens with two perpendicular joint sets. Based on the test results, they established four distinct failure modes, namely, splitting or tensile failure of intact blocks, shearing of intact blocks, rotation of intact blocks and sliding along the joints. These failure modes are linked to the joint orientations.

To obtain realistic mechanical properties of jointed rock masses under various joint orientations, a number of large volumes of rock mass configurations should be tested at high stresses under different stress paths (Kulatilake et al. 2015). In situ tests can provide direct information on the strength and deformability of rock masses; however, as noted by Bieniawski (1973), it is difficult to rely on one in situ test alone for engineering design because different results may be obtained even in a fairly uniform and good quality rock mass. Therefore, investigating the effects of joint orientation on the mechanical properties of a large volume jointed rock mass using these direct measurements is not only difficult but also time-consuming and very expensive.

At the laboratory scale, although laboratory testing using small-sized rock specimens containing discontinuities cannot provide accurate rock mass properties, they can still be used to investigate the failure modes and study how strength and deformation properties of jointed rock masses are affected by the number of joint sets or joint orientations.

Alejano et al. (2017) proposed a novel method that combines the HB failure criterion and laboratory test results based on standard laboratory-size rock specimens with two joint sets to estimate the strength of jointed rock masses. They suggested that properties of hard rock masses are more significantly linked to rock structures than to sample scales. Small laboratory-size specimens with a structure similar to large-scale rock masses could provide useful information on rock mass behavior at the engineering scale.

To better understand the fundamental failure mechanisms of blocky rock masses and to improve the accuracy of rock strength estimation of empirical models based on the GSI system, UDEC rock mass models, which are calibrated by the laboratory data, are used in this research to investigate the effects of joint orientation on the strength of blocky rock mass with two joint sets in the unconfined state. Based on the analysis of numerical results, a modified UCS model is proposed, which is capable of giving more accurate estimation of UCS of jointed rock masses with various joint orientations.

9.2 Methods for Estimating Rock Mass Strength Using Laboratory-Size Specimens

9.2.1 Rock Specimen Configuration

Arzúa et al. (2014) and Alejano et al. (2017) carried out extensive uniaxial and triaxial compression tests on intact rocks and jointed rock masses to investigate the strength and deformation responses of jointed granitic rock specimens with two joint sets.

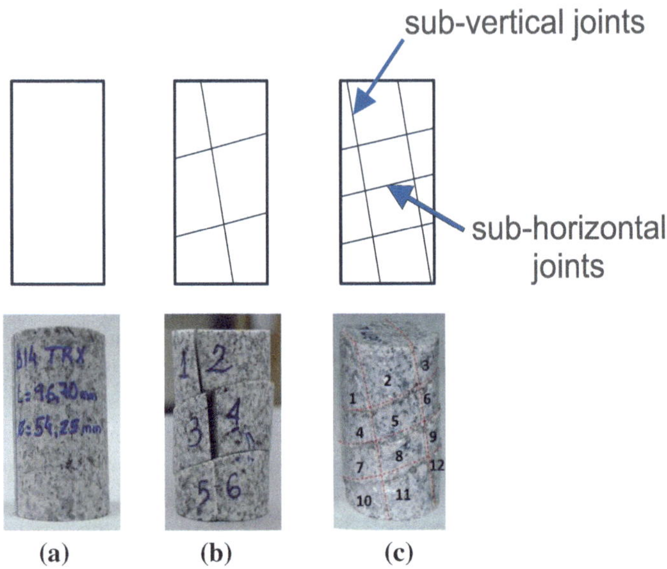

Fig. 9.1 Rock specimens used for laboratory tests: **a** intact rock specimen, **b** 1 + 2 jointed rock mass specimen, **c** 2 + 3 jointed rock mass specimen (Alejano et al. 2017)

The tested rock is bright, white-colored and coarse-grained granite. All the specimens tested came from the same homogeneous granite block. Three different types of rock specimens with the mean diameter of 54.0 mm and the mean height of 96.7 mm, were tested and they are: intact rocks (Fig. 9.1a), 1 + 2 jointed specimens (Fig. 9.1b) and 2 + 3 jointed specimens (Fig. 9.1c). The joint surfaces are planar and smooth with a slightly rough surface due to sawing.

9.2.2 *Quantifying the Degree of Fracturing Using Equivalent GSI*

In order to build the relationships between the strengths of laboratory-size jointed specimens and the HB failure criterion which is usually used for estimating rock mass strength at the engineering scale, a concept of equivalent GSI was proposed by Alejano et al. (2017) to quantify the degree of fracturing of laboratory-sized rock specimens. Alejano et al. (2017) suggested that small laboratory-size specimens with a structure homothetic to large-scale rock masses could provide useful information on the characteristic of rock mass behavior at the engineering scale.

The strategy to calculate the equivalent GSI for rock specimens is to use the HB failure criterion to fit triaxial data (Fig. 9.2) using the nonlinear regression method.

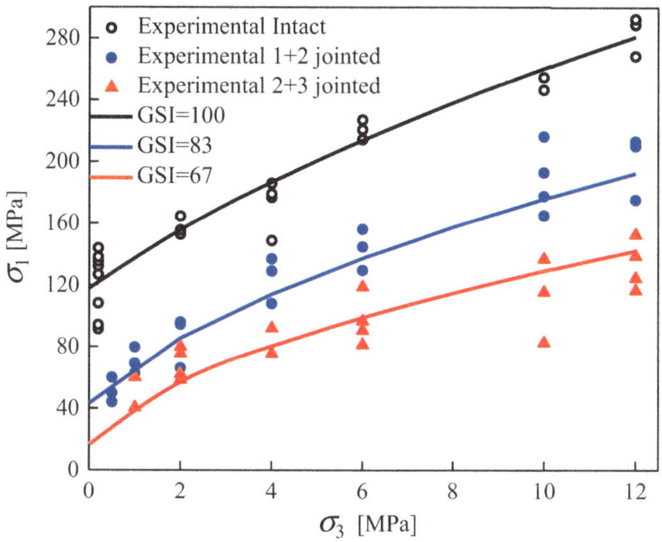

Fig. 9.2 Triaxial compression test results of intact and jointed rock specimens (data from Arzúa et al. 2014; Alejano et al. 2017)

Substituting Eqs. (9.2) to (9.4) into Eq. (9.1), the HB failure criterion can be expressed as $\sigma_1 = f$ (GSI, m_i, σ_{ci}, D and σ_3). The values of σ_{ci} = 123.35 MPa and m_i = 41.69 is estimated from the uniaxial and triaxial compression tests of the intact rock. The disturbance factor D is assumed to be zero. Therefore, the HB failure envelopes that give the best fit to the triaxial test data (σ_1, σ_3) can be generated by adjusting the values of GSI. The final fitting results are shown in Fig. 9.2 which give GSI = 100 for the intact rock, GSI = 83 for the 1 + 2 jointed specimen and GSI = 67 for the 2 + 3 jointed specimen.

9.3 Calibration of UDEC Rock Mass Models

In this research, the Universal Distinct Element Code (UDEC) was used to build numerical rock mass models to investigate the mechanical behavior of jointed rock masses under compression. UDEC is a two-dimensional (2D) discontinuum modeling program based on the distinct element method that can simulate the responses of jointed rock masses when subjected to either quasi-static or dynamic loading. UDEC is widely used in rock engineering for applications ranging from studies of the progressive failure of rock materials (Alshkane et al. 2017) to evaluations of the influences of rock joints on the stability of slopes and underground excavations (He et al. 2018; Zheng et al. 2018b).

In a UDEC model, a rock mass is represented as an assemblage of discrete blocks. The discontinuities are modelled using special boundary conditions between blocks so large displacements along discontinuities and rotations of blocks are permitted. UDEC has several built-in material constitutive models, for both intact blocks and discontinuities, which allow the simulation of mechanical behaviors representative of discontinuous geological (or similar) materials. In this research, individual blocks behave as deformable materials. These deformable blocks are sub-divided into a mesh of elements, and the failure criterion of each element follows the HB criterion. For discontinuities, the relative motions in normal and shear directions are governed by the Coulomb-slip force-displacement relationship (Itasca 2011).

For most of the nonlinear block models in UDEC (such as the HB model), plastic zone can be plotted when the stresses of elements satisfy the yield criterion. Such an indication usually represents that plastic flow is occurring, but it is possible for block zones simply to "sit" on the yield surface without any significant flow taking place. Therefore, it is important to look at the whole pattern of plasticity indicators to see if a mechanism which is indicated if there is a contiguous line of active plastic zones that join two surfaces has developed. The diagnosis is confirmed if the velocity plot also indicates motion corresponding to the same mechanism. Note that only the actively yielding elements ("at yield surface", as shown in Fig. 9.7) are important to the detection of a failure mechanism (Itasca 2011).

The configuration of jointed rock mass models with two perpendicular joint sets is shown in Fig. 9.3.

In this research, the central cross section of the 3D rock mass model (Fig. 9.3a) was used for 2D simulation analyses, as shown in Fig. 9.3b. The model has plain stress elements with non zero constant stress in the plane. In other words, the value of σ_2 which is perpendicular to the plane keeps the same as confining pressure σ_3 during the loading process of triaxial tests.

The 2D model has a width of 54.0 mm, a height of 96.7 mm and a joint spacing ratio of $S_1/S_2 = 1$. β is the acute angle between joint set 1 and the major principal stress direction. Intact rocks are modeled with elastic-perfectly plastic constitutive behavior using the Hoek–Brown failure criterion to ensure that the UDEC analysis can capture shear and tensile failures within the intact blocks. The Coulomb-slip model (also elastic-perfectly plastic) is used to simulate the joint strength and deformation behavior. The axial strain loading rate is set to 10 mm/s for uniaxial and triaxial compression test simulations.

9.3.1 Calibration for Intact Rock Model

In order to test the reliability of the UDEC model for intact rocks, triaxial compression tests with confining stresses ranging from 0 to 12 MPa were carried out. The input parameters for the intact rock model are: uniaxial compressive strength $\sigma_{ci} = 123.35$ MPa, Young's modulus $E = 27.95$ GPa, Poisson's ratio $\nu = 0.17$,

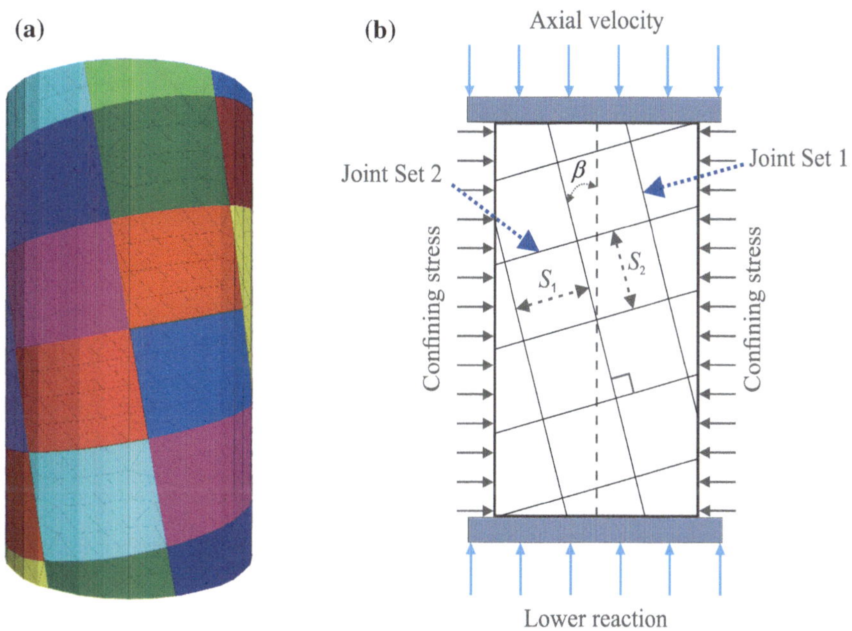

Fig. 9.3 Configuration of the blocky rock mass model **a** 3D rock mass model, **b** 2D rock mass model which is the central cross section of the 3D model

m_i = 41.69, and unit weight γ = 23 kN/m^3. Stress–strain curves under 4 MPa confinement obtained from the laboratory test and the UDEC simulation are shown in Fig. 9.4. It can be seen that peak strength obtained from the numerical simulation is very close to the laboratory test result. The results presented in Fig. 9.5 show that the triaxial strength obtained from the UDEC simulation agrees well with the experimental data.

9.3.2 Calibration for Rock Mass Models

In order to build reliable rock mass models that can reflect similar mechanical behavior of rock masses with that obtained from laboratory tests, there is a need to identify the joint properties for a selected constitutive model. In this research, the Coulomb-slip model is used to simulate joint strength and deformation behavior. The model has five input parameters: joint normal stiffness k_n, shear stiffness k_s, joint tensile strength σ_{tj}, joint cohesion c_j and joint friction ϕ_j. A parametric study was carried out to identify the joint parameters to ensure that the UDEC model can generate the same response as that from the laboratory tests by Arzúa et al. (2014) and Alejano et al. (2017).

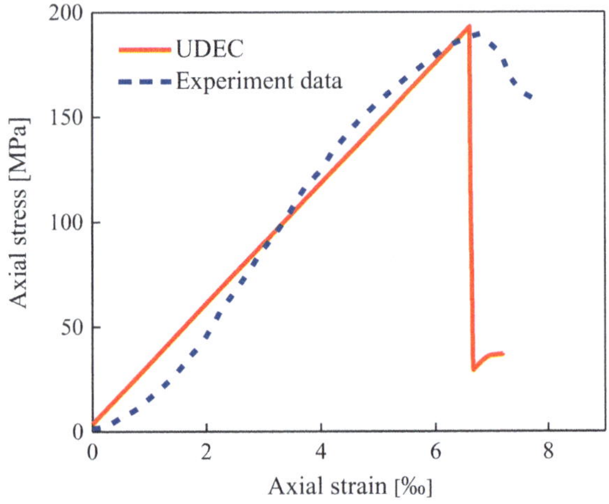

Fig. 9.4 Comparison of stress-strain curves (σ_3 = 4 MPa)

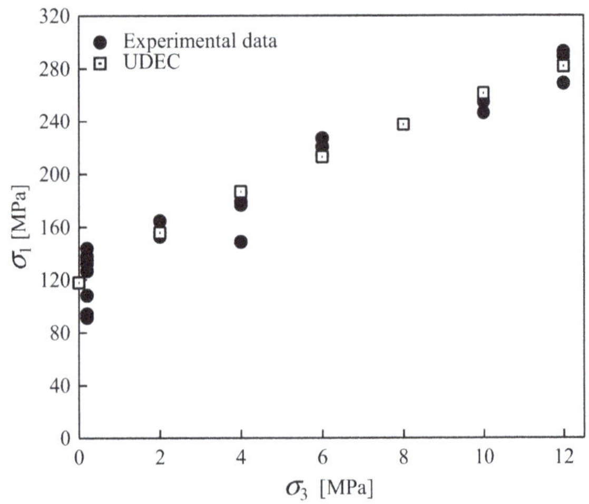

Fig. 9.5 Comparison of triaxial strengths obtained from experiments and numerical simulations for intact rocks

Considering that the joint surfaces are planar and smooth and slightly rough due to sawing, σ_{tj} = 0 MPa, c_j = 0 MPa and $\phi_j = 40°$ are assumed in the numerical model. The selection of k_n and k_s is based on the inverse-modeling calibration approach. It is known that the deformations of a joint in the normal and tangential directions affect the global behavior of a rock mass. Vergara et al. (2016) suggested that the k_n/k_s ratio should be between 2 and 3, therefore, 2 is used in this study. Based on our parameter studies, the final values of k_n = 60 GPa/m, k_s = 30 GPa/m and $\phi_j = 40°$ were determined for the UDEC model.

Figure 9.6 compares the numerical simulation results with the experimental data, showing that the strength predicted by the numerical model for the 1 + 2 jointed specimens are in good agreement with the experimental results. For the 2 + 3

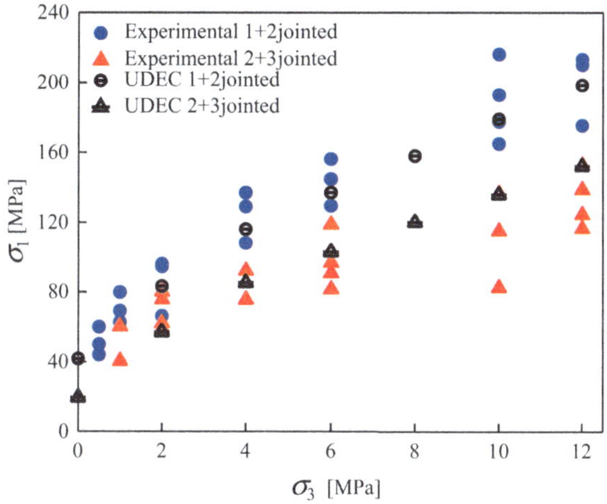

Fig. 9.6 Comparison of strength of jointed rock at different confining stress calculated in UDEC using plain stress elements versus triaxial strength for jointed granite measured at the lab

jointed specimens, the numerical results are good in the low confinement zone but slightly over-estimate the rock mass strength in the high confinement zone ($\sigma_3 > 6$ MPa).

Figure 9.7 shows the failure patterns of a model under triaxial compression at three deformation stages. The specimen undergoes elastic deformation, crack initiation, stable crack propagation, unstable crack propagation and crack coalescence, with intact block failure as the load increases. Shear bands are observed in the top

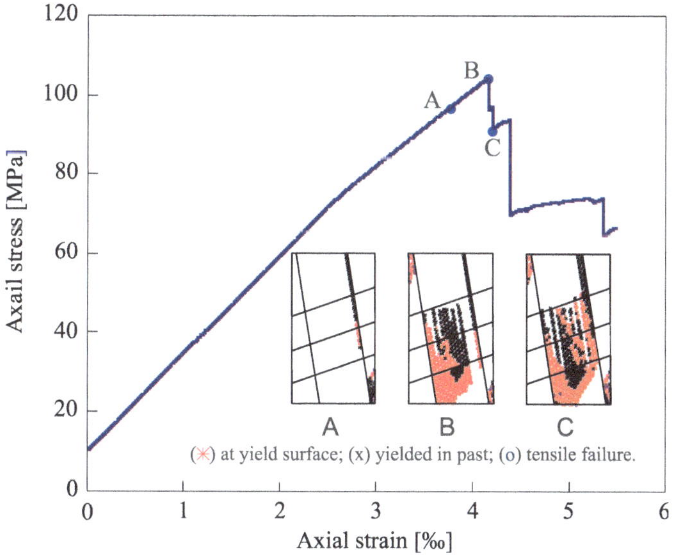

Fig. 9.7 Stress–strain curve and failure process for the 2 + 3 jointed model under 10 MPa confining pressure

and bottom areas of the model, following pre-existing joints (Stage A). Failure in intact block is typically due to newly formed shear bands or axial splitting cracks (Stage B). The post-peak failure is brittle. Tensile failures occur continuously and they are parallel with the sub-vertical joints; shear bands continue develop in the bottom of the model (Stage C).

The results presented in Figs. 9.4, 9.5, 9.6 and 9.7 indicate that the UDEC models with the selected intact rock and joint properties can capture reasonable failure behavior of rock masses with two joint sets, which means that they can be used for rock mass strength estimation.

It should be noted that 2D UDEC model used in this research cannot generate the same stress and strain distributions of rock samples as 3D experimental tests. This is a common known intrinsic limitation of the 2D software. However 2D analyses still have the capability to reveal the fundamental failure mechanisms of rock masses and are widely used in rock engineering (Fan et al. 2013; Shen et al. 2013; Wu and Fan 2014; Wu et al. 2017; Zhang and Wong 2018; Zheng et al. 2018a; Meng et al. 2019).

Here, we carried out a comparison study between 2D and 3D numerical simulations for the 1 + 2 jointed rock mass sample to show that the use of 2D models can also capture the same fundamental rock failure patterns as that of 3D models. The 3D analyses were carried out using three-dimensional Distinct Element Code (3DEC) software (Itasca 2016). The 3DEC rock mass model has the same intact and joint properties as the 2D model.

Figure 9.8 shows that shear bands are observed in the top-left and bottom-right areas for both the 3DEC and UDEC models. Simulation results show that the peak strength obtained from 3DEC and UDEC are 192.76 and 198.50 MPa, which is close to the average peak strength obtained from the laboratory tests that is 199.72 MPa. The discrepancies for the 3DEC and UDEC models are −3.48 and −0.61%, respectively. As input parameters of joints are calibrated based on the UDEC model as described in Sect. 9.3.2, therefore, it is not surprise that the UDEC model gives more accurate peak strength than that of the 3DEC model.

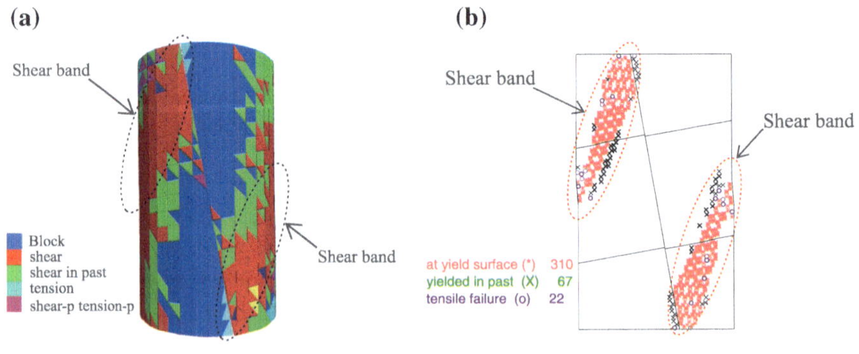

Fig. 9.8 Failure patterns for the 1 + 2 jointed rock mass model under triaxial tests with confining pressures 12 MPa, **a** failure pattern of the 3DEC model, **b** failure pattern of the UDEC model

9.4 Rock Mass Configurations for Numerical Simulations

Once the UDEC models are validated, they can then be used to investigate the mechanical behavior of jointed rock masses with various joint orientations under different stress conditions. In this fashion, a database can be generated to supplement the data obtained from the existing laboratory tests based on which a modified empirical UCS model for anisotropic rock masses is developed.

Two additional rock mass models, which have two perpendicular joint sets, were added for the numerical analyses. The two models are named as 3 + 5 jointed and 5 + 7 jointed rock mass models (see Fig. 9.9) and they have the same intact rock and joint properties as the 1 + 2 jointed or 2 + 3 jointed rock mass models but different joint spacing as shown in Table 9.1. Numerical triaxial compression tests were carried out to obtain the strength of the additional models and the results are shown in Fig. 9.10.

The strategy introduced in Sect. 9.2 is used to derive the equivalent GSI values for the two models, which are 40 and 34 for the 3 + 5 and 5 + 7 jointed rock mass models, respectively.

Figure 9.11 shows a series of rock mass models with different joint orientations ranging from 0° to 90° at an interval of 10°. The jointed rock mass models, which are symmetrical with respect to $\beta = 45°$, are used to investigate the influence of joint orientation on rock mass strength and failure mode in an unconfined state.

9.5 Numerical Simulation Results

As mentioned above, in the UDEC rock mass model, the joint cohesion c_j is assumed to be zero. Therefore, some specimens will fail under their own dead-weights at certain joint orientations. In order to deal with this problem, a small

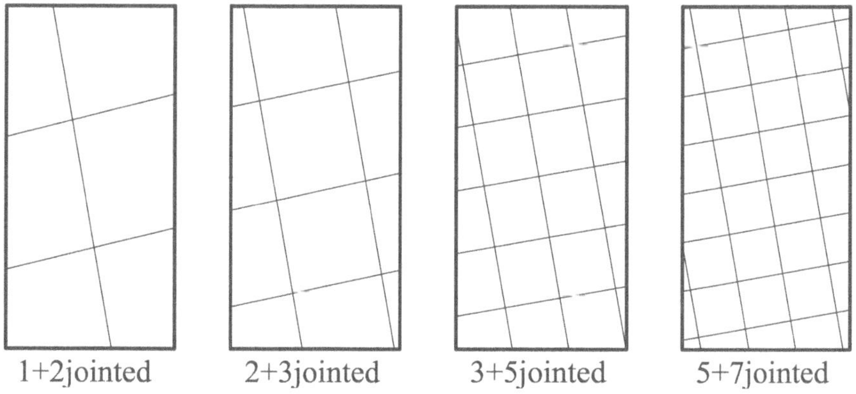

Fig. 9.9 Various rock mass models used for numerical simulations

Table 9.1 The spacing and equivalent GSI values of rock models

Model	Intact	1 + 2 jointed	2 + 3 jointed	3 + 5 jointed	5 + 7 jointed
GSI	100	83	67	40	34
Spacing (mm)	–	40	30	20	15

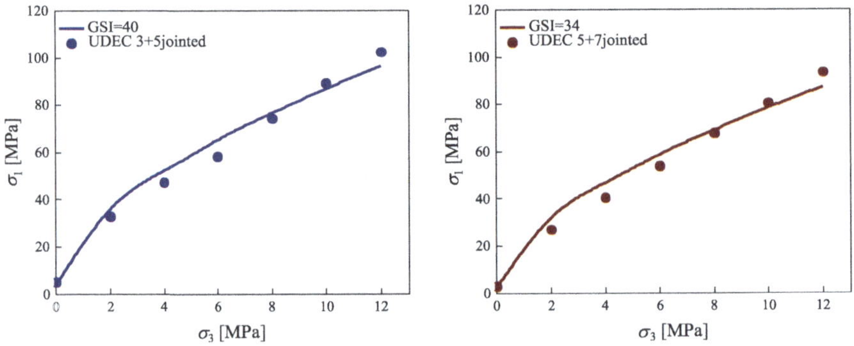

Fig. 9.10 GSI-equivalent results for **a** 3 + 5 and **b** 5 + 7 jointed rock mass models

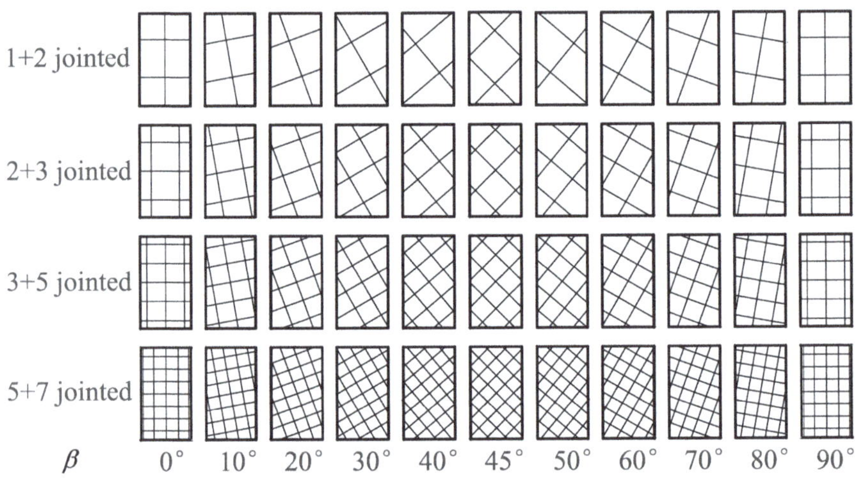

Fig. 9.11 Jointed rock mass models for further numerical simulations

confining stress $\sigma_3 = 0.1$ MPa was applied in the UCS tests. This technique was also adopted by Brown and Trollope (1970) in their physical model tests. The comprehensive numerical results of UCS and failure modes for the jointed rock mass models are summarized in Tables 9.2 and 9.3. Results presented in Table 9.2 are also plotted in Fig. 9.12. Results in Fig. 9.12 show that the UCS decreases when β increases from 0° to 40°. However, the strength increases abruptly due to a

Table 9.2 UCS (unit: MPa) of jointed rock masses at different joint orientations

β (°)	1 + 2 jointed	2 + 3 jointed	3 + 5 jointed	5 + 7 jointed
0	95.10	85.55	76.91	69.11
10	46.62	19.90	5.58	3.06
20	8.41	4.82	1.77	1.47
30	2.86	2.43	1.06	0.93
40	3.44	2.43	1.37	1.27
45	8.73	6.35	1.88	1.49
50	3.44	2.43	1.37	1.27
60	2.86	2.43	1.06	0.93
70	8.41	4.82	1.77	1.47
80	46.62	19.90	5.58	3.06
90	95.10	85.55	76.91	69.11

change of failure mode from sliding to rotation at $\beta = 45°$. At this angle, the intersection of the two joint sets forms two perfectly aligned touching block tips that can sustain a reasonable amount of vertical load, resulting in such an outcome. Similar behavior was also observed in Ghazvinian and Hadei (2012).

The effects of joint orientations on the failure pattern of rock masses under low confining stress conditions were reported in Singh et al. (2002). They defined four typical modes of rock mass failure, namely, splitting (or tensile), shearing, sliding and rotation failures. In this research, each jointed rock mass model consists of intact rock blocks and joints. Both shear and tensile failures are possible within the intact rock blocks, and sliding and rotation failures are possible for the joints. Detailed observations of various failure modes based on the modeling results are summarized below.

Shear failure mode: the main mechanism of failure for models having joint orientation $\beta = 0°$ is shear failure within intact rock blocks. This mode of failure involves the appearance of shear bands within the blocks, and the subsequent dilation of the model in the minor principal stress direction. These models exhibit very high strengths (see Fig. 9.12).

Tensile failure mode: for models having joint orientation of $\beta = 10°–20°$, the main mechanism of failure is splitting or tensile failure within the intact rock blocks. The models exhibit relatively high compressive strengths. Tensile fractures parallel to the direction of the sub-vertical joints are observed in these models, as shown in second and third rows in Table 9.3. These fractures occur when the effective tensile stress reaches the tensile strength of the intact rock. For the 3 + 5 and 5 + 7 jointed rock mass models with lower GSI values, the failure mode changes to the combination of tensile and sliding failures. No macroscopic movements parallel or normal to the planes of the pre-existing joints are observed.

Sliding failure mode: The main mechanism of failure for models having joint orientation of $\beta = 30°–40°$ is sliding through pre-existing joints. The mode is

Table 9.3 Failure patterns of jointed rock masses at different joint orientations

β(°)	1+2 jointed	2+3 jointed	3+5 jointed	5+7 jointed
0	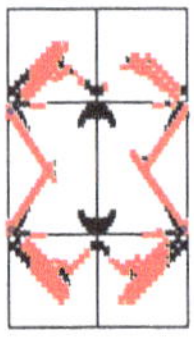		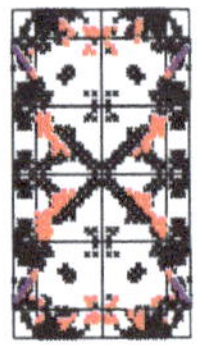	
10		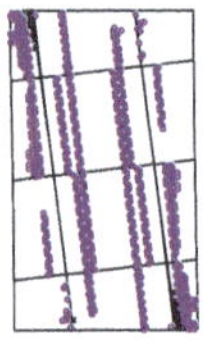	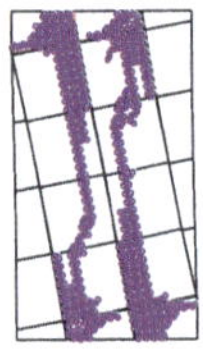	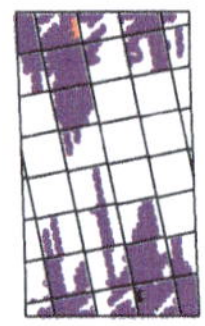
20	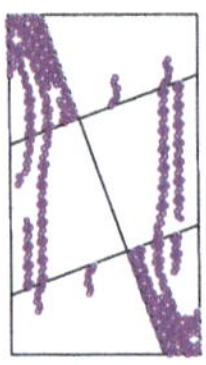	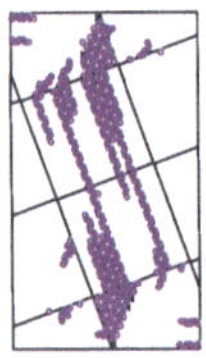	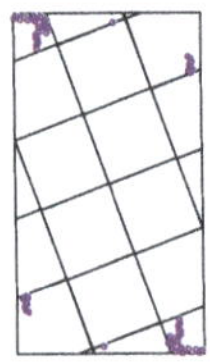	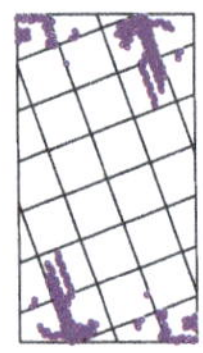
30		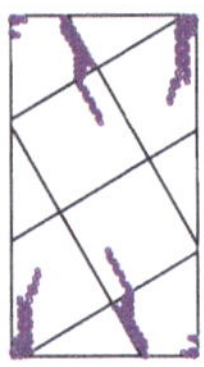		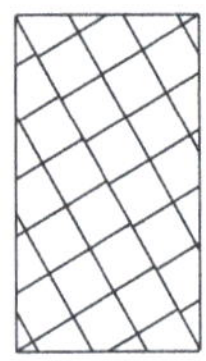
40				
45			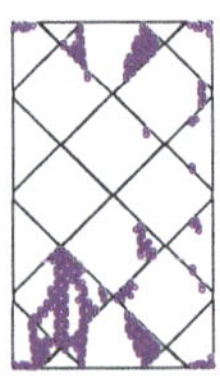	

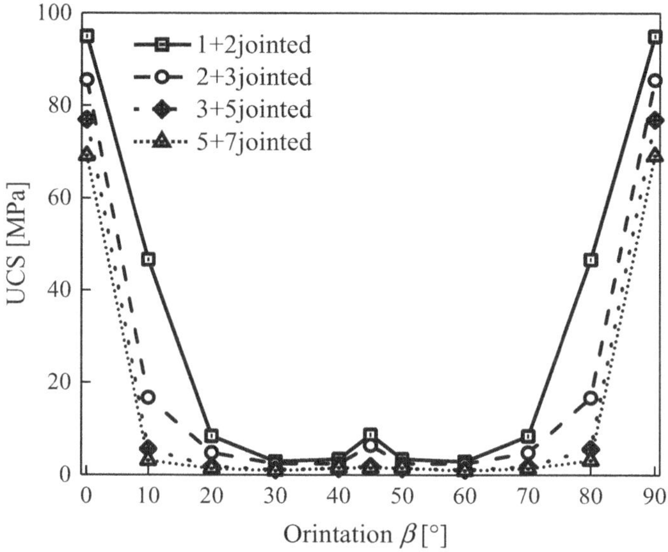

Fig. 9.12 Relationship between the UCS of jointed rock masses and joint orientations

associated with large deformations and a poorly defined peak in the stress–strain curves. Note that the basic friction angle of joints is 40°. This mode is sometimes associated with intact rock tensile failure for the 2 + 3 jointed rock mass models.

Rotation failure mode: for models having a joint dip of 45°, the main mechanism of failure is rotation failure. Under this failure mode, rotation of intact rock blocks takes place from the beginning of loading. The specimen as a whole translates and large relative displacement in the transverse direction is observed. For the 2 + 3, 3 + 5 and 5 + 7 jointed rock mass models, failure occurs within the intact rock blocks partially due to tensile failure and partially due to rotation.

These failure modes for various jointed rock mass models presented in Table 9.3 are summarized in Table 9.4.

Table 9.4 Main failure modes for rock mass models at different joint orientations

β (°)	1 + 2 jointed	2 + 3 jointed	3 + 5 jointed	5 + 7 jointed
0	Shear			
10	Tensile			
20				
30	Slide			
40				
45	Rotation			

9.6 Modified Anisotropic UCS Model for Jointed Rock Masses

9.6.1 Comparison of UCS Obtained from Empirical and Numerical Models

The empirical UCS model based on the HB failure criterion can be written as:

$$\sigma_{\text{cm}} = \sigma_{\text{ci}} s^{a} \tag{9.5}$$

where σ_{cm} is the UCS of the rock mass, s and a are the HB input parameters that can be estimated from GSI and D using Eqs. (9.3) and (9.4).

As mentioned above, the GSI system does not consider the effect of joint orientation on the quality of a rock mass. Therefore, Eq. (9.5), which is based on the GSI system, has to be used with caution for blocky rock masses with two joint sets, as shown in Fig. 9.11. The numerical simulation results presented in Sect. 9.5 clearly show that the strength and failure modes of a jointed rock mass can be significantly influenced by joint orientations, especially when $\beta < 10°$.

Therefore, the existing empirical equation to estimate UCS of jointed rock masses based on the GSI system should be modified to incorporate the effect of joint orientation so that they can be used for estimating anisotropic rock mass strength. Based on the discussions above, the UCS values obtained from experiments and numerical simulations as well as empirical equations are summarized in Table 9.5 and Fig. 9.13.

It should be noted that the values of $\sigma_{\text{cm-Lab}}$ for jointed rock masses are calculated from triaxial compression test data using the fitted curve, as shown in Fig. 9.2, because Arzúa et al. (2014) and Alejano et al. (2017) did not carry out UCS test on jointed rock specimens directly. $\sigma_{\text{cm-E}}$ is calculated from Eq. (9.5) using the equivalent GSI value of the rock specimens (D is assumed to be zero in this case). $\sigma_{\text{cm\_N}}$ is the UCS of the numerical rock mass model with a joint orientation of $\beta = 10°$, because when $\beta = 10°$, the UCS obtained from the UDEC model is very close to the value from the empirical model [Eq. (9.5)]. Therefore, the rock mass strength model for $\beta = 10°$ is regarded as the reference model that can be used for direct comparison of results between laboratory test and empirical models.

Table 9.5 Comparison of UCS (unit: MPa) obtained from different methods

Models	1 + 2 jointed	2 + 3 jointed	3 + 5 jointed	5 + 7 jointed
GSI	83	67	40	34
$\sigma_{\text{cm\_N}}$, numerical ($\beta = 10°$)	46.62	19.90	5.27	3.07
$\sigma_{\text{cm\_E}}$, empirical	47.14	19.08	4.08	2.78
$\sigma_{\text{cm\_L}}$, laboratory	46.30	19.69	–	–

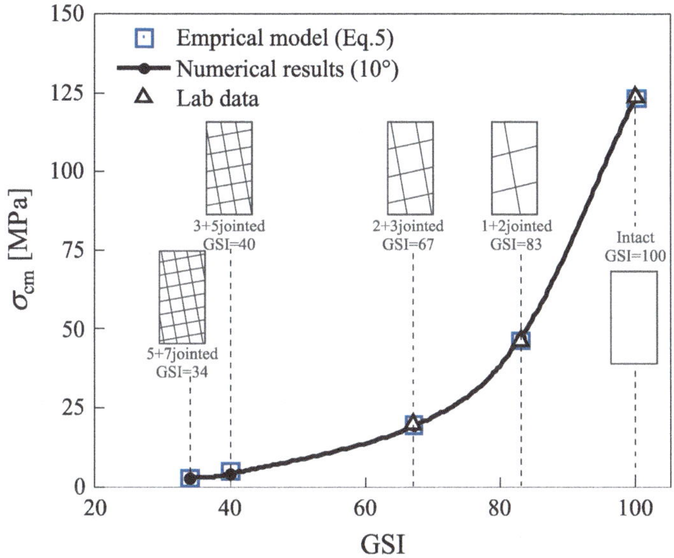

Fig. 9.13 Relationship between the UCS and GSI derived from jointed rock mass models when $\beta = 10°$

9.6.2 *Anisotropic Weighting Factor*

To quantify the effects of joint orientation on UCS of jointed rock masses, an anisotropic weighting factor f_β is proposed as:

$$f_\beta = \frac{\sigma_{cm\_N(\beta)}}{\sigma_{cm\_N(10^\circ)}} \tag{9.6}$$

where $\sigma_{cm\_N(\beta)}$ is the UCS of a jointed rock mass with joint orientation β, and $\sigma_{cm\_N(10^\circ)}$ is the UCS of a rock mass at $\beta = 10°$. Based on the results listed in Table 9.2, the values of f_β for different joint orientations can be calculated and the results are shown in Table 9.6 and plotted in Fig. 9.14.

We also investigated the effects of joint friction j_ϕ on the anisotropic weighting factor f_β, comprehensive simulation results are shown in Table 9.7 and Fig. 9.15.

Table 9.6 Values of f_β for various rock mass models

β (°)	0	10	20	30	40	45	50	60	70	80	90
1 + 2 jointed	2.04	1.00	0.18	0.06	0.07	0.19	0.07	0.06	0.18	1.00	2.04
2 + 3 jointed	5.12	1.00	0.29	0.15	0.15	0.38	0.15	0.15	0.29	1.00	5.12
3 + 5 jointed	13.78	1.00	0.32	0.19	0.25	0.34	0.25	0.19	0.32	1.00	13.78
5 + 7 jointed	22.50	1.00	0.46	0.30	0.48	0.49	0.48	0.30	0.46	1.00	22.50

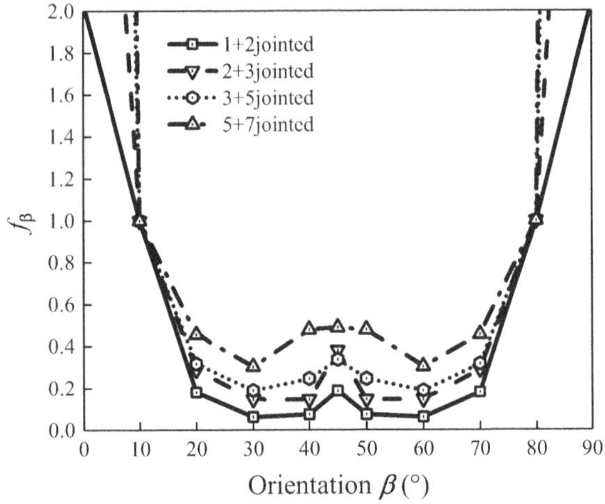

Fig. 9.14 Values of f_β for different joint orientations

Table 9.7 Effects of j_ϕ on f_β for 1 + 2 jointed rock mass model

j_ϕ(°)	β = 10°	β = 20°	β = 30°	β = 40°	β = 45°
0	F	F	F	F	F
10	1.00	0.15	F	F	0.16
20	1.00	0.18	F	F	0.17
30	1.00	0.18	0.05	0.07	0.18
40	1.00	0.18	0.06	0.07	0.19
50	1.00	0.19	0.08	0.07	0.19
60	1.00	0.20	0.12	0.07	0.20
70	1.00	0.33	1.54	0.53	0.55
80	1.00	0.63	1.32	0.40	0.45
95	1.00	0.60	0.89	0.25	0.28

F means the rock mass model fails due to its own deadweight

It is found that the value of anisotropic weighting factor f_β will be significantly affected by the values of joint friction when $j_\phi > 60°$. This is because of the increase of joint friction at some point the rock mass model would stop sliding along the joint or rotating of intact rock blocks, that means the failure mode changes from slide or rotation failures to split or shear failures, such an example is shown in Fig. 9.16. It should be noted that in practical the joint friction is generally less than 60°, in such condition, the effect of joint friction on f_β is relatively small comparing with the high values of joint friction, as shown in Fig. 9.15.

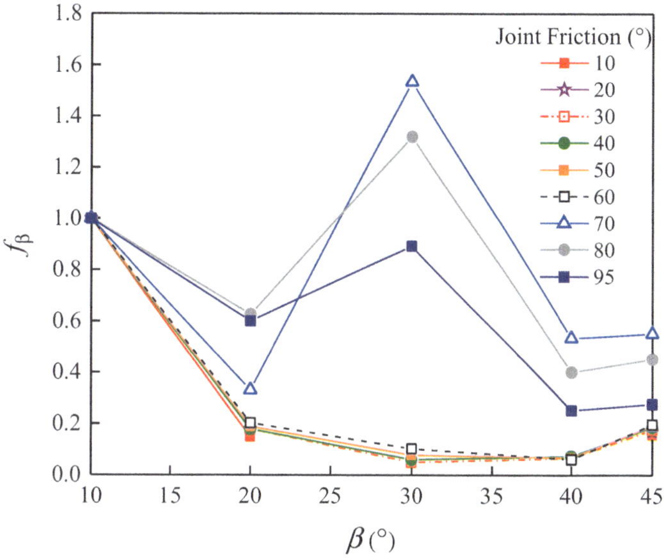

Fig. 9.15 Effects of joint friction on f_β for the 1 + 2 jointed sample

Fig. 9.16 Failure pattern changes with the increase of joint friction for 1 + 2 jointed rock mass model, **a** slide failure with $j_\phi = 60°$, **b** split failure with $j_\phi = 95°$

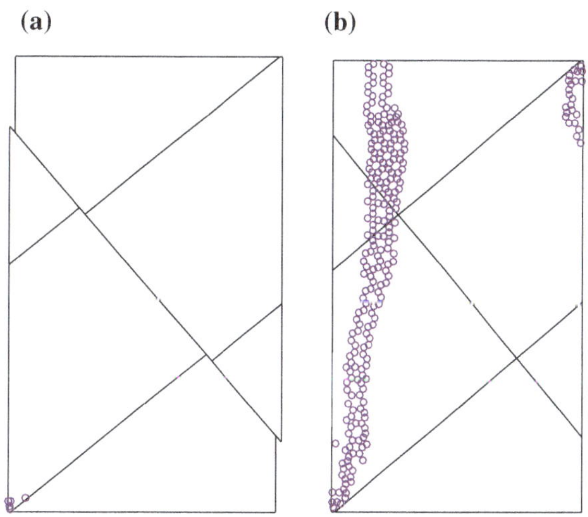

9.6.3 The Modified UCS Model

As concluded by Alejano et al. (2017), small samples with a structure homothetic to large-scale rock masses could provide useful guidelines on rock mass behaviors at the engineering scale. The values of f_β derived based on numerical simulations of laboratory-size models are believed to be also applicable to large-scale rock masses.

Therefore, the following modified UCS model is proposed for the estimation of anisotropic rock mass strength considering joint orientation:

$$\sigma_{\rm cm} = f_{\beta}\sigma_{\rm ci}s^{a} \tag{9.7}$$

As shown in Fig. 9.14, $f_{\beta\leftarrow} > 1$ when $\beta < 10°$, meaning that the existing model of Eq. (9.5) is conservative for design. However, when $10° < \beta < 45°$, $f_{\beta\leftarrow} < 1$, meaning that the actual UCS of a rock mass is grossly overestimated by Eq. (9.5), which can result in an unsafe engineering design. Based on the results shown in Fig. 9.17, a non-linear equation [Eq. (9.8)] is used to describe the relation between f_β and β for $f_{\beta\leftarrow} < 1$, with joint orientations between 10° and 80°:

$$f_\beta = A(|\beta - 45°| - B)^2 + C \tag{9.8}$$

where A, B and C are constants related to the GSI value of a rock mass, and their values can be obtained from regression analyses shown in Fig. 9.18. Note the relation is symmetrical with respect to $\beta = 45°$. The shapes of the f_β curves estimated from Eq. (9.8) for different rock mass configurations are similar to these shown in Fig. 9.17.

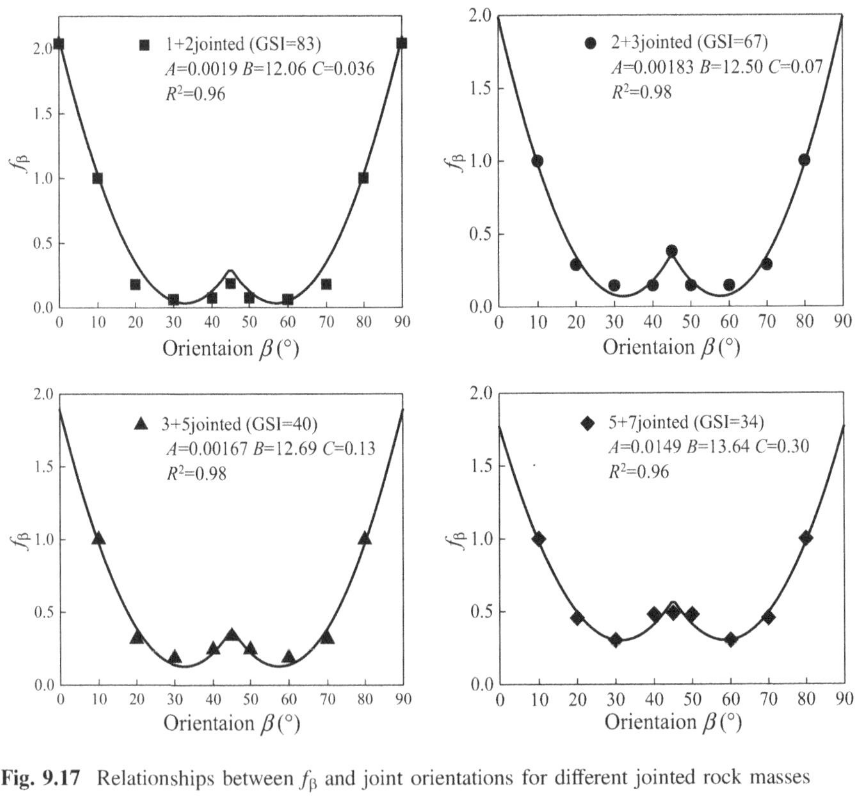

Fig. 9.17 Relationships between f_β and joint orientations for different jointed rock masses

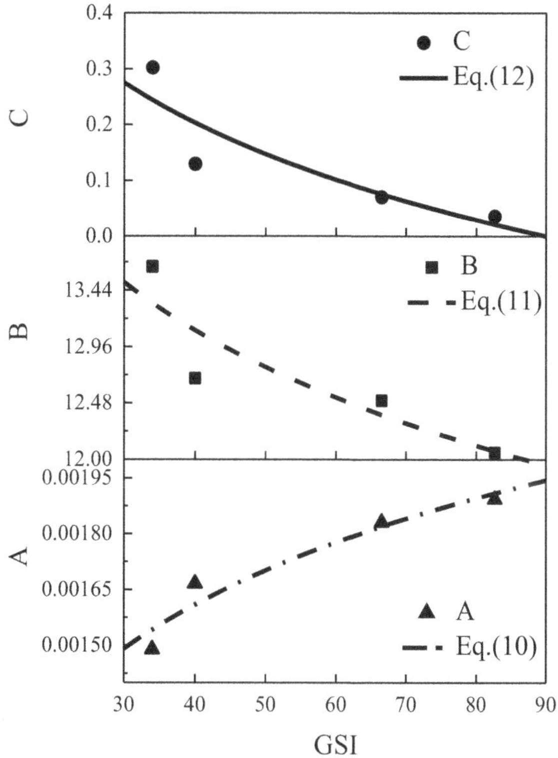

Fig. 9.18 Relationships between GSI and fitting constants A, B and C

Finally, the modified empirical UCS model can be expressed as:

$$\sigma_{\mathrm{cm}} = \left(A(|\beta - 45^{\circ}| - B)^2 + C\right)\sigma_{\mathrm{ci}} s^a \tag{9.9}$$

$$A = 4.13 \times 10^{-4}\ \ln(GSI) + 8.71 \times 10^{-5} \tag{9.10}$$

$$B = -1.42\ln(GSI) + 18.32 \tag{9.11}$$

$$C = -0.25\ln(GSI) + 1.13 \tag{9.12}$$

Comparing with the traditional model [Eq. (9.5)], the proposed empirical UCS model is capable of giving a better estimate of UCS for rock masses with different joint orientations. This is important for engineering applications because traditional models can grossly overestimate the UCS of some rock masses which may impose a risk on engineering design.

9.7 Conclusions

In this research, UDEC rock mass models, calibrated using published experimental data, were used to investigate the effects of joint orientation β on the strength of jointed rock masses with two joint sets in the unconfined state. The model is capable of capturing the mechanical behaviors of jointed rock masses obtained from laboratory tests.

The UCSs obtained from the UDEC simulations were compared with those calculated from the traditional empirical UCS model [Eq. (9.5)] which is based on the GSI system that does not incorporate the effects of joint orientation on the strength of rock mass. It is shown that the traditional model [Eq. (9.5)] is conservative in the estimation of the rock mass strength for engineering design when $\beta < 10°$. However, the values of UCS for $10° < \beta < 45°$ are grossly overestimated by the traditional model, which can impose risk in engineering design.

To rectify the problem, based on the analysis of the numerical simulation results, an anisotropic weighting factor f_β was proposed and incorporated into Eq. (9.5) for the estimation of the anisotropic UCS of rock masses with different joint orientations. The proposed empirical UCS model can provide better estimations of anisotropic rock mass strengths due to the effect of joint orientation, compared with the traditional model, which can result in safer engineering design.

References

Alejano LR, Arzúa J, Bozorgzadeh N, Harrison JP (2017) Triaxial strength and deformability of intact and increasingly jointed granite samples. Int J Rock Mech Min Sci 95:87–103

Alshkane YM, Marshall AM, Stace LR (2017) Prediction of strength and deformability of an interlocked blocky rock mass using UDEC. J Rock Mech Geotech Eng 9(3):149–160

Arzúa J, Alejano LR, Walton G (2014) Strength and dilation of jointed granite specimens in servo-controlled triaxial tests. Int J Rock Mech Min Sci 69(3):93–104

Bieniawski ZT (1973) Engineering classification of jointed rock masses. Trans S Afr Inst Civ Eng 15(12):335–343

Brown ET, Trollope DH (1970) Strength of a model of jointed rock. J Soil Mech Found Div 96:685–704

Cai M, Kaiser PK, Uno H, Tasaka Y, Minami M (2004) Estimation of rock mass deformation modulus and strength of jointed hard rock masses using the GSI system. Int J Rock Mech Min Sci 41(1):3–19

Cao R, Cao P, Fan X, Xiong X, Lin H (2016) An experimental and numerical study on mechanical behavior of ubiquitous-joint brittle rock-like specimens under uniaxial compression. Rock Mech Rock Eng 49(11):1–20

Darlington WJ, Ranjith PG, Choi SK (2011) The effect of specimen size on strength and other properties in laboratory testing of rock and rock-like cementitious brittle materials. Rock Mech Rock Eng 44(5):513–529

Dershowitz WS, Einstein HH (1988) Characterizing rock joint geometry with joint system models. Rock Mech Rock Eng 21(1):21–51

Fan L, Yi X, Ma G (2013) Numerical manifold method (NMM) simulation of stress wave propagation through fractured rock mass. Int J Appl Mech 5(2):1350022

Ghazvinian A, Hadei MR (2012) Effect of discontinuity orientation and confinement on the strength of jointed anisotropic rocks. Int J Rock Mech Min Sci 55(10):117–124

He S, Li Y, Aydin A (2018) A comparative study of UDEC simulations of an unsupported rock tunnel. Tunn Undergr Space Technol 72:242–249

Hoek E, Brown ET (1980) Underground excavations in rock. The Institution of Mining and Metallurgy, London

Hoek E, Brown ET (2018) The Hoek-Brown failure criterion and GSI—2018 edition. J Rock Mech Geotech Eng. https://doi.org/10.1016/j.jrmge.2018.08.001

Hoek E, Carranza-Torres C, Corkum B (2002) Hoek-Brown failure criterion. In: Bawden HRW, Curran J, Telesnicki M (eds) Proceedings of the North American rock mechanics society (NARMS-TAC 2002), 2002 edition. Mining Innovation and Technology, Toronto, Canada, pp 267–273

Hoek E, Carter T, Diederichs M (2013) Quantification of the geological strength index chart. In: 47th US rock mechanics/geomechanics symposium, San Francisco: ARMA13-672

Itasca (2011) UDEC (universal distinct element code). Version 5.0. Minneapolis

Itasca (2016) 3DEC (three-dimensional distinct element code). Version 5.20. Minneapolis

Jimenez R, Serrano A, Olalla C (2008) Linearization of the Hoek and Brown rock failure criterion for tunnelling in elasto-plastic rock masses. Int J Rock Mech Min Sci 45(7):1153–1163

Kim B, Cai M, Kaiser PK, Yang H (2007) Estimation of block sizes for rock masses with non-persistent joints. Rock Mech Rock Eng 40(2):169–192

Kulatilake PHSW, Liang J, Gao H (2001) Experimental and numerical simulations of jointed rock block strength under uniaxial loading. J Eng Mech-ASCE 127(12):1240–1247

Kulatilake PHSW, He W, Um J, Wang H (2015) A physical model study of jointed rock mass strength under uniaxial compressive loading. Int J Rock Mech Min Sci 34(3–4):165.e1–165.e15

Meng J, Cao P, Huang J, Lin H, Chen Y, Cao R (2019) Second-order cone programming formulation of discontinuous deformation analysis. Int J Numer Methods Eng. https://doi.org/10.1002/nme.6006

Shen J, Karakus M (2014) Three-dimensional numerical analysis for rock slope stability using shear strength reduction method. Can Geotech J 51(2):164–172

Shen J, Priest S, Karakus M (2012) Determination of Mohr-Coulomb shear strength parameters from generalized Hoek-Brown criterion for slope stability analysis. Rock Mech Rock Eng 45:123–129

Shen J, Karakus M, Xu C (2013) Chart-based slope stability assessment using the generalized Hoek-Brown criterion. Int J Rock Mech Min Sci 64:210–219

Singh M, Rao KS, Ramamurthy T (2002) Strength and deformational behaviour of a jointed rock mass. Rock Mech Rock Eng 35(1):45–64

Vásárhelyi B, Kovács L, Ákos T (2016) Analysing the modified Hoek-Brown failure criteria using Hungarian granitic rocks. Geomech Geophys Geo-Energy Geo-Resour 2(2):131–136

Vergara MR, Jan MVS, Lorig L (2016) Numerical model for the study of the strength and failure modes of rock containing non-persistent joints. Rock Mech Rock Eng 49(4):1211–1226

Vorobiev OY, Rubin MB (2018) A thermomechanical anisotropic continuum model for geological materials with multiple joint sets. Int J Numer Anal Methods Geomech 42:1366–1388

Wang W, Shen J (2017) Comparison of existing methods and a new tensile strength based model in estimating the Hoek-Brown constant mi for intact rocks. Eng Geol 224:87–96

Wang Y, Jing H, Su H, Xie J (2017) Effect of a fault fracture zone on the stability of tunnel-surrounding rock. Int J Geomech 17(6):1–20

Wang Y, Guo P, Dai F, Li X, Zhao Y, Liu Y (2018) Behavior and modeling of fiber-reinforced clay under triaxial compression by combining the superposition method with the energy-based homogenization technique. Int J Geomech 18(12):04018172

Wu Z, Fan L (2014) The numerical manifold method for elastic wave propagation in rock with time-dependent absorbing boundary conditions. Eng Anal Bound Elements 46:41–50

Wu Z, Fan L, Liu Q, Ma G (2017) Micro-mechanical modeling of the macro-mechanical response and fracture behavior of rock using the numerical manifold method. Eng Geol 225:49–60

Yang Z, Chen J, Huang T (1998) Effect of joint sets on the strength and deformation of rock mass models. Int J Rock Mech Min Sci 35(1):75–84

Zhang Y, Wong L (2018) A review of numerical techniques approaching microstructures of crystalline rocks. Comput Geotech 115:167–187

Zhang Y, Ren F, Yang T, Wang S, Zhang W, Yu M (2018) An improved rock mass characterization method using a quantified geological strength index and synthetic rock mass model. Rock Mech Rock Eng 51(11):3521–3536

Zheng H, Li T, Shen J, Xu C, Sun H, Lü Q (2018a) The effects of blast damage zone thickness on rock slope stability. Eng Geol 246:19–27

Zheng Y, Chen C, Liu T, Zhang H, Xia K, Liu F (2018b) Study on the mechanisms of flexural toppling failure in anti-inclined rock slopes using numerical and limit equilibrium models. Eng Geol 237:116–128

Zuo J, Liu H, Li H (2015) A theoretical derivation of the Hoek-Brown failure criterion for rock materials. J Rock Mech Geotech Eng 7(4):361–366

Chapter 10
Non-linear Shear Strength Reduction Method for Slope Stability Based on the HB Criterion

Abstract Existing numerical modeling of three-dimensional (3D) slopes is mainly performed by the shear strength reduction (SSR) technique based on the linear Mohr-Coulomb (MC) criterion, whereas the non-linear failure criterion for rock slope stability is seldom used in slope modeling. However, it is known that rock mass strength is a non-linear stress function and that, therefore, the linear MC criterion does not agree with the rock mass failure envelope very well. In this research, therefore, a non-linear SSR technique is proposed that can use the Hoek-Brown (HB) criterion to represent the non-linear behavior of a rock mass in $FLAC^{3D}$ program to analyze 3D slope stability. Extensive case studies are carried out to investigate the influence of convergence criterion and boundary conditions on the 3D slope modeling. Results show that the convergence criterion used in the 3D model plays an important role, not only in terms of the calculation of the factor of safety (FOS), but also in terms of the shape of the failure surface. The case studies also demonstrate that the value of the FOS for a given slope will be significantly influenced by the boundary condition when the slope angle is less than 50°.

10.1 Introduction

Rock slope stability is one of the major challenges of rock engineering projects, such as open pit mining. Rock slope failure can affect mining operations and result in costly losses in terms of time and productivity. Therefore, the evaluation of the stability of rock slopes is a critical component of open pit design and operation (Naghadehi et al. 2013).

In most of the geotechnical applications two-dimensional (2D) plain strain analysis is commonly used to simulate stability of earth structures (Basarir et al. 2005; Karakus et al. 2007; Kurakus 2007; Eid 2010; Tutluoglu et al. 2011). The majority of rock slope analyses in practical projects are still performed using 2D limit equilibrium or plane strain analysis because the 2D analysis is relatively simple and yields a conservative factor of safety (FOS) compared with

J. Zuo and J. Shen, *The Hoek-Brown Failure criterion—From theory to application*,
https://doi.org/10.1007/978-981-15-1769-3_10

Table 10.1 3D slope stability analysis using different methods

Authors	Methods
Cheng and Yip (2007)	LEM
Griffiths and Marquez (2007)	SSR
Frazaneh et al. (2008)	LAM
Li et al. (2009)	LAM
Michalowski and Drescher (2009)	LAM
Wei et al. (2009)	SSR/LEM
Li et al. (2010)	LAM
Michalowski (2010)	LAM
Detournay et al. (2011)	SSR
Stianson et al. (2011)	SSR
Gharti et al. (2012)	SSR
Zheng (2012)	LEM
Nian et al. (2012)	SSR
Michalowski and Nadukuru (2013)	LAM
Nadukuru and Michalowski (2013)	LAM
Zhang et al. (2013)	SSR

three-dimensional (3D) analysis (Griffiths and Marquez 2007). Please refer to comparative studies between 2D and 3D slope analysis conducted by Li et al. (2009), Michalowski (2010) and Zhang et al. (2013). However, it is known that 3D analysis provides the more realistic model because it can take into account the appropriate geometry and boundary conditions. Therefore, the development of 3D slope analysis has become a popular research topic in geotechnical engineering in recent years. A list of 3D slope stability papers published is shown in Table 10.1.

Commonly used approaches for 3D slope stability analysis include: the limit equilibrium method (LEM), limit analysis method (LAM), and numerical modelings, such as the finite element method (FEM) and discrete element method (DEM). FEM is derived from the continuum theory and has the merit of simulating continuous bodies. DEM describes particulate materials and is particularly well suited to analyze jointed media and multi-body interaction (Shen and Abbas 2013). The 3D LEM model involves various assumptions about the internal force distribution, and it is difficult to locate the critical failure surface, as is well documented in the literature (Griffiths and Marquez 2007; Wei et al. 2009; Zhang et al. 2013). The 3D LAM model has been used for slopes with simple geometries. However, the construction of the 3D failure mechanism for LAM is not straightforward for complicated slope models, which leads to this method being seldom used for complex conditions (Wei et al. 2009).

Currently, 3D numerical modeling performed by the shear strength reduction (SSR) technique is a very attractive and commonly accepted approach among geotechnical researchers and engineers because it not only can automatically locate the critical failure surface, but can also simulate the stress-strain behavior and give the progressive shear failure of the slope in complex geometries and loading conditions.

Although the SSR technique has the above merits, the authors still must take into account its limitations in order to use it for the analysis of 3D isotropic rock slopes, as follows: (1) the existing 3D SSR technique is mainly based on the linear Mohr-Coulomb (MC) criterion. It is known that rock strength is non-linear, and many researches (Priest 2005; Li et al. 2008; Jimenez et al. 2008; Shen et al. 2012a; Tao et al. 2013) showed that the MC criterion generally cannot represent rock mass behavior very well, especially for slope stability problems where the rock mass is in a state of low confining stresses that make the non-linearity more obvious; (2) the selection of appropriate convergence criterion is not easy for a 3D SSR model because the value of the FOS for a given slope can be considerably influenced by the convergence criterion; (3) boundary conditions play an important role in the distribution of internal stresses in the slope model and can affect the simulation results.

With the aim of better understanding the fundamental rock slope failure mechanisms and improving the accuracy of the rock slope stability results of 3D numerical models, in this research a simple non-linear SSR technique is proposed to be used with the Hoek-Brown (HB) criterion which can ideally represent the non-linear behavior of a rock mass, in FLAC^{3D} program (Itasca 2009) in order to analyze 3D slope stability. Extensive case studies are carried out to investigate the influence of the convergence criterion and boundary conditions on the numerical results which include rock mass shear strength, the shape of the failure surface, as well as the FOS values.

10.2 Instantaneous Shear Strength of the HB Criterion

The non-linear HB criterion, initially proposed by Hoek and Brown (1980), has been widely used for predicting intact rock and rock mass strength in rock engineering for several decades. The latest version of the HB criterion presented by Hoek et al. (2002) is expressed as:

$$\sigma_1 = \sigma_3 + \sigma_{ci}\left(m_b \frac{\sigma_3}{\sigma_{ci}} + s\right)^a \tag{10.1}$$

where σ_1 and σ_3 are the maximum and minimum principal stresses, σ_{ci} is the uniaxial compressive strength (UCS) of the intact rock. m_b, s and a are the Hoek-Brown input parameters which can be estimated from the Geological Strength Index (GSI), disturbance factor D and intact rock constant m_i.

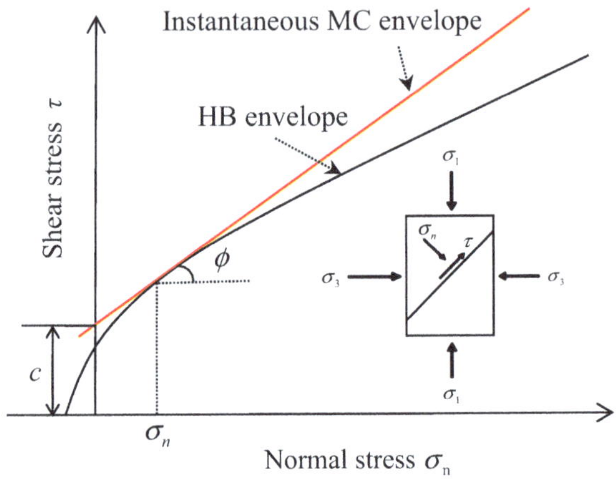

Fig. 10.1 Instantaneous MC envelope of the HB criterion in the normal and shear stress plane

$$m_b = m_i e^{\left(\frac{GSI-100}{28-14D}\right)} \tag{10.2}$$

$$s = e^{\left(\frac{GSI-100}{9-3D}\right)} \tag{10.3}$$

$$a = 0.5 + \frac{e^{\left(-\frac{GSI}{15}\right)} - e^{\left(\frac{-20}{3}\right)}}{6} \tag{10.4}$$

In order to use the HB criterion in conjunction with SSR methods for calculating the FOS of rock slopes, methods are required to determine the instantaneous MC shear strength parameters of cohesion c and angle of friction ϕ from the HB criterion (Fu and Liao 2010). The HB criterion (see Eq. 10.1) is expressed by the relationship between maximum and minimum principal stresses. However, it can also be expressed in terms of normal stress $\sigma_{\rm n}$ and shear stress τ on the failure plane as shown in Fig. 10.1. The instantaneous cohesion c and angle of friction ϕ can be calculated by locating the tangent of the HB envelope under a given value of normal stress $\sigma_{\rm n}$, as illustrated in Fig. 10.1. The intercept with the τ axis gives the c value, and the slope of the tangent to the HB failure envelope yields the ϕ value.

Figure 10.1 also illustrates the stress state of an element where the strength can be defined by the MC criterion. If the stress state (σ_1, σ_3) of an element is known, the corresponding instantaneous c and ϕ values can be calculated using Eqs. 9.5–9.8 proposed by Shen et al. (2012b).

$$\frac{\sigma_n}{\sigma_{ci}} = \frac{\sigma_3}{\sigma_{ci}} + \frac{\left(\frac{m_b\sigma_3}{\sigma_{ci}} + s\right)^a}{2 + am_b\left(\frac{m_b\sigma_3}{\sigma_{ci}} + s\right)^{a-1}} \tag{10.5}$$

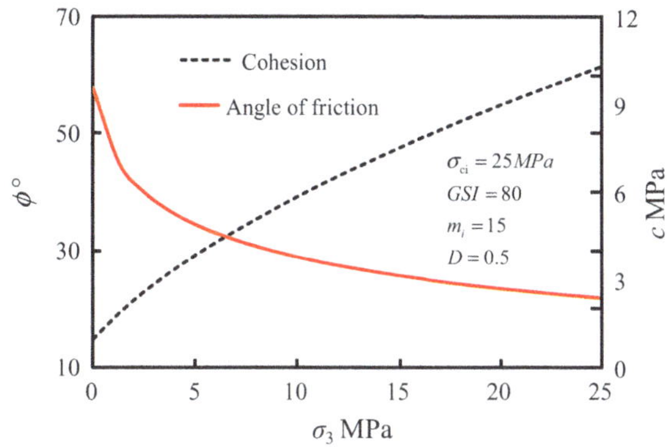

Fig. 10.2 The correlations between MC parameters and σ_3

$$\phi = \arcsin\left(1 - \frac{2}{2 + am_b\left(m_b\frac{\sigma_3}{\sigma_{ci}} + s\right)^{a-1}}\right) \tag{10.6}$$

$$\tau = \frac{\sigma_{ci}\cos\phi}{2\left(1 + \frac{\sin\phi}{a}\right)^a}\left(m_b\frac{\sigma_n}{\sigma_{ci}} + s\right)^a \tag{10.7}$$

$$c = \tau - \sigma_n \tan\phi \tag{10.8}$$

The numerical slope model can be divided into numbers of elements using mesh techniques. When the slope is modeled under the loading condition, the stress states of the elements in the model will vary, which leads to the elements having different values of c and ϕ.

An example can be used to show the relationship between instantaneous c, ϕ and minimum principal stress σ_3, as shown in Fig. 10.2. The following parameters were used for the calculation: σ_{ci} = 25 MPa, GSI = 80, m_i = 15, D = 0.5; the values of σ_3 range from 0 to 25 MPa. Figure 10.2 illustrates that the values of instantaneous c increase and ϕ decrease with the increase of σ_3 values, which reflects the non-linear behavior of the HB criterion

10.3 Non-linear SSR Method for the HB Criterion

The calculation of the FOS using the SSR technique is based on reducing the MC shear strength parameters c and ϕ until the slope failures, and then the value of the FOS can be defined as the ratio of the actual shear strength to the minimum shear strength of the rock or soil materials that are required to prevent failure (Duncan 1996). The reduced shear strength parameters c_f and ϕ_f are given by:

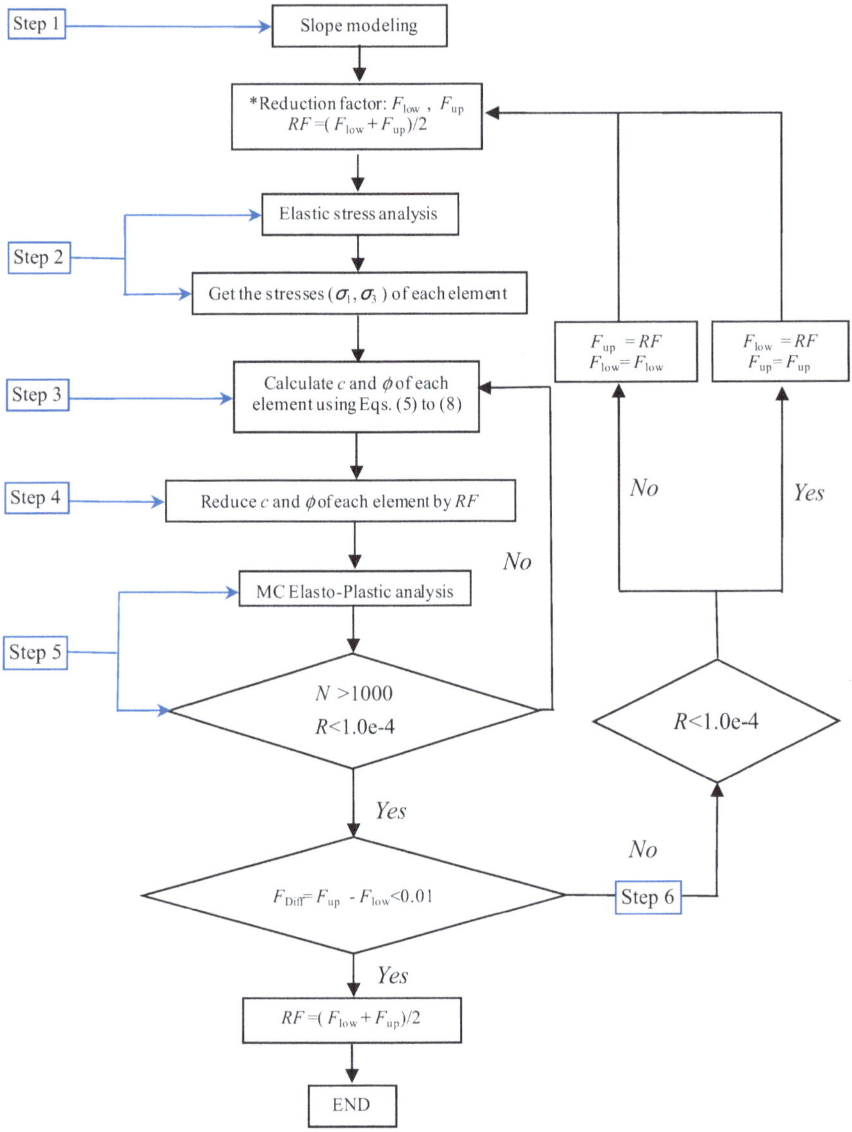

Fig. 10.3 Flow chart of the application of HB criterion into FLAC3D using the non-linear SSR technique. * The values of RF can be adjusted using the bracketing approach proposed by Dawson et al. (1999). F_{low} and F_{up} are the lower and upper bracket values of FOS. F_{Diff} is the difference between upper and lower FOS values

$$c_f = \frac{c}{RF} \tag{10.9}$$

$$\phi_f = \arctan\frac{\tan\phi}{RF} \tag{10.10}$$

where RF is a reduction factor, and the value of RF is equal to the FOS when slope failure occurs.

One of the most promising ways to use the HB criterion in conjunction with SSR techniques is to estimate the instantaneous MC shear strength parameters c and ϕ for elements in the slope. The details of the application of this non-linear SSR technique for 2D FEM slope analysis can be found in the paper by Fu and Liao (2010). Kumar's (1998) solution was used by Fu and Liao (2010) to calculate the instantaneous ϕ values, which requires Newton's iteration formula to calculate the ϕ values. It should be noted that Eqs. 10.5–10.8 are an alternative form of Kumar's (1998) solution. However, the equations offer the benefit of being able to calculate the instantaneous c and ϕ of an element from its' stress state (σ_1, σ_3) without the need of iteration analysis.

The non-linear SSR strategy, together with Eqs. 10.5–10.8, was used to implement the HB criterion in FLAC^{3D} for 3D rock slope stability analysis. Figure 10.3 is a flow chart showing the steps of implementing the HB criterion in the FLAC^{3D} slope model, as follows: Step 1: Build the slope model according to slope geometry, rock mass properties, loading and boundary conditions. Mesh techniques are used to generate the grid elements for a slope. Step 2: Carry out the elastic stress analysis to determine the stress state of each element in the slope model. Step 3: Use Eqs. 10.5–10.8 to calculate the cohesion c as well as the angle of friction ϕ for each element. Step 4: Reduce the c and ϕ values of all elements by a reduction factor RF. Step 5: Use the reduced c_f and ϕ_f for the elasto-plastic analysis using the MC constitutive model. Step 6: Repeat steps 2 to 5 when a new reduction factor RF is generated until slope failure. Finally, the value of the FOS of a given slope is equal to the reduction factor RF.

10.4 Convergence Criterion in the 3D Models

Research by Wei et al. (2009) demonstrated that the value of the factor of safety of the 3D SSR model can be significantly influenced by the selection of the convergence criterion. Therefore, it is necessary to carry out some trial and error analysis to select an appropriate convergence criterion for a slope model.

The convergence criterion in FLAC^{3D} is the nodal unbalanced force ratio R, and the user must specify the number of calculation steps N to bring the model to a state of equilibrium. An example of data from the paper by Hammah et al. (2005) can be used to check the influence of the convergence criterion on the 3D slope model. The example has the following slope geometry and rock mass properties: slope height

H = 10 m, slope angle $\beta = 45°$, σ_{ci} = 30 MPa, m_i = 2, GSI = 5, D = 0, unit weight γ = 25 kN/m^3, Deformation modulus E_m = 5000 MPa and Poisson's ratio v = 0.3.

The model has 475 elements and the analyses were carried out using 1 m unit width. The boundary conditions for the slope are: *x*-direction displacement at the front and back faces of the slope model are fixed; *x, y* and *z*-direction displacement at the base face of the slope model are fixed; and *y*-direction displacement of the end faces of the slope model are fixed (see Fig. 10.4).

The results show that when the convergence criteria change, the values of the FOS vary from 1.01 to 1.90. Calculation step *N* has a slight effect on the FOS when *N* is more than 1000. For example, when *R* = 1E-3, the value of the FOS is equal to 1.82 for *N* = 1000 and FOS increases to 1.90 for *N* = 2000. The results also indicate a clear trend in the increase of the FOS with the increase of *R*. For example, when *R* values are increased from 1E-5 to 1E-4, *N* = 1000, the values of the FOS increase from 1.04 to 1.29, where the difference is 19.4%. In this case, FOS = 1.29 is more realistic than FOS = 1.04 as the model cannot form the failure surface when the value of *R* is equal to 1E-5. Eight slope cases were analyzed, which have the same parameters as the slope from Hammah et al. (2005), with various slope

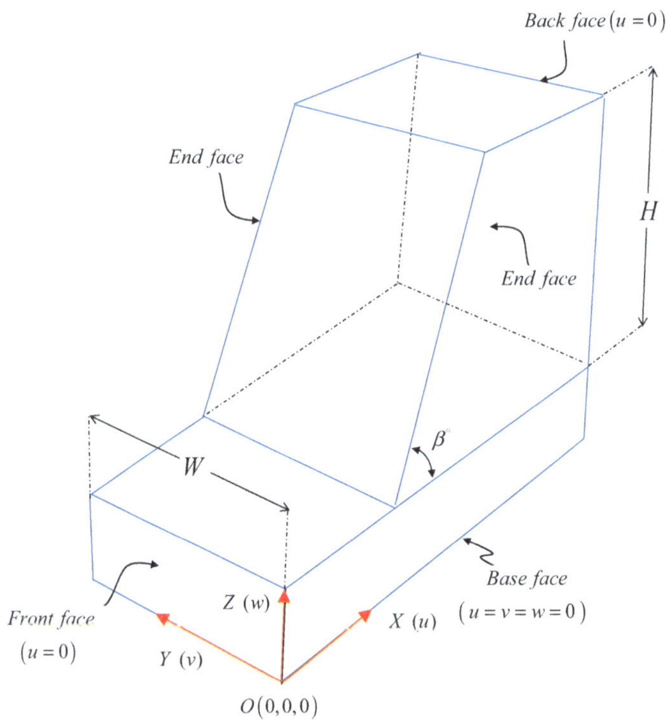

Fig. 10.4 Boundary conditions for a slope model

geometries ($\beta = 60°$ and $70°$, $H = 5$ m and 20 m) and rock properties ($\sigma_{ci} = 20$ MPa and 50 MPa, $D = 0.3$ and 0.5). It was found that the 8 slope models cannot produce a failure surface when R = 1E-5. However, the model will produce the failure surface, and the FOS value tends to stabilize when the convergence criterion is R = 1E-4 and the value of N is more than 1000.

Figure 10.5 compares the failure surfaces corresponding to the FOS values using various convergence criteria in the slope model. N is the iteration number and R is the convergence parameter.

The choice of the mesh techniques can also influence the FOS results in the SSR analysis. A comprehensive study of the influence of different mesh techniques and mesh elements on the calculation of the FOS using FLAC^{3D} models has been conducted by Zhang et al. (2013). Figure 10.6 compares the different mesh sizes for the calculation of the FOS corresponding to the computation time. The results show that when the mesh size is more than 1400, the value of the FOS tends to equal 1.16, which is close to Hammah et al.'s (2005) result of a FOS of 1.15.

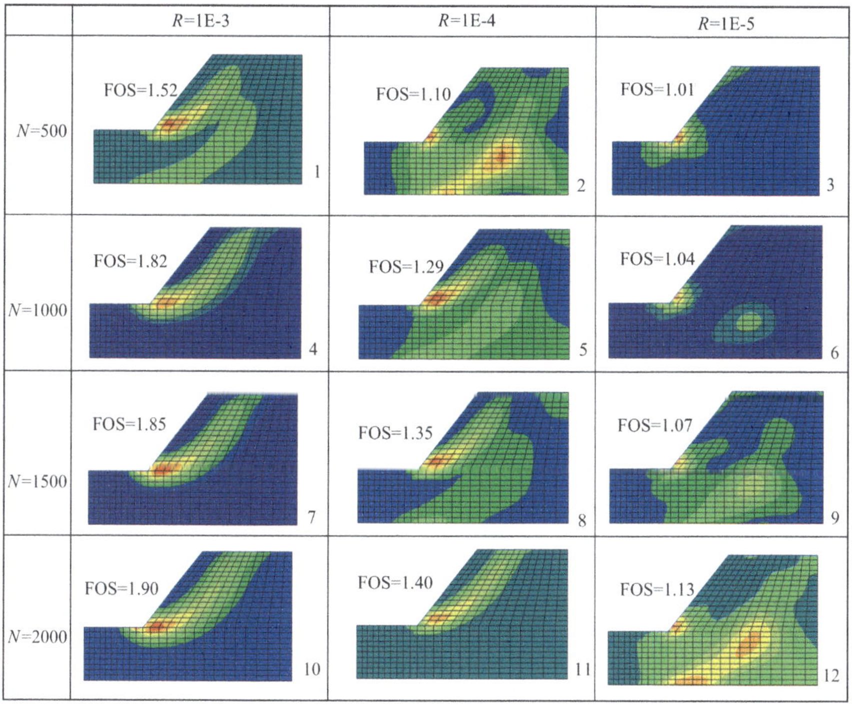

Fig. 10.5 Comparison of failure surfaces corresponding to FOS values using different convergence criteria (N is the iteration number and R is the convergence parameter)

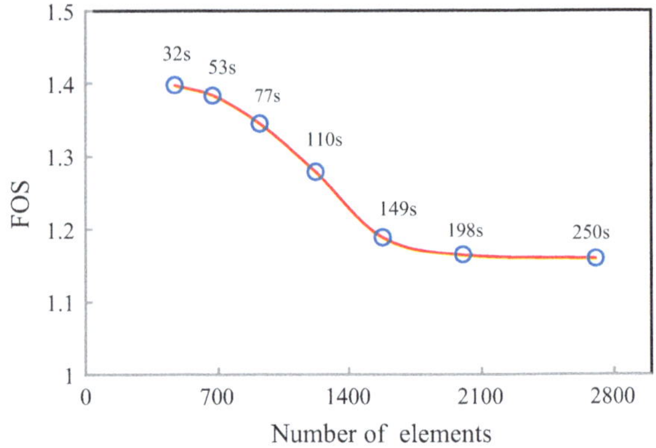

Fig. 10.6 Plot of FOS values corresponding to computation time versus mesh elements (On a computer with Intel (R) Core Duo, 2.53 GHz)

10.5 Boundary Conditions in 3D Models

The choice of appropriate boundary conditions is important for 3D slope stability analysis as boundary conditions play an important role in the development of internal stresses in a slope, which will change the shape of failure surface corresponding to the value of the FOS.

The commonly used boundary conditions for a 3D slope model (see Fig. 10.4) are: fixing the x-direction displacement ($u = 0$) at the front and back faces of slope model; fixing the x, y and z-direction displacement ($v = u = w = 0$) at the base of slope model. For the end faces, there are three types of boundary conditions as suggested by Chugh (2003). Condition 1: fixing the y-direction displacement ($v = 0$), which represents contact with a rigid, smooth abutment that can provide a reacting thrust but no in-plane shear restraint; Condition 2: fixing the x, y-direction displacement ($u = v = 0$), which represents a contact that provides side shear resistance; Condition 3: fixing the x, y, z-direction displacement ($u = v = w = 0$), which is used to characterize contacts with no movement.

A small-scale slope (Table 10.2) with modest rock mass damage under good blasting ($D = 0.7$) was used to analyze the influence of boundary conditions on slope stability. Figure 10.7 compares the slope failure surfaces corresponding to the

Table 10.2 Input parameters of a slope case

Input parameters	Values
H, m	20
β, °	60
γ, kN/m^3	27
σ_{ci}, MPa	5
GSI	40
m_i	12
D	0.7

Boundary condition used at the end faces of the 3D models			
	Condition 1 (Fix y)	Condition 2 (Fix x, y)	Condition 3 (Fix x, y, z)
Failure surface	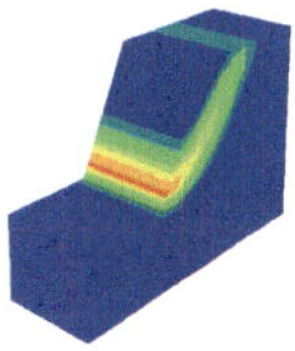	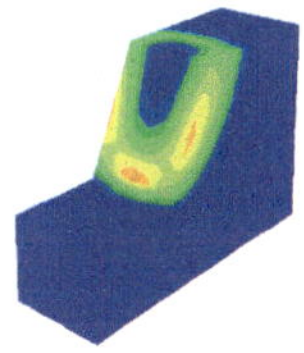	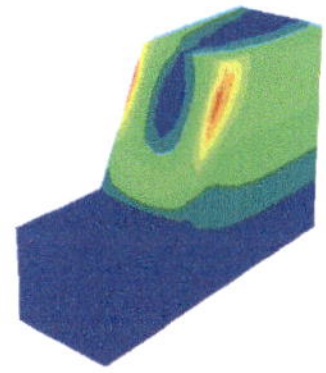
Contour of Cohesion	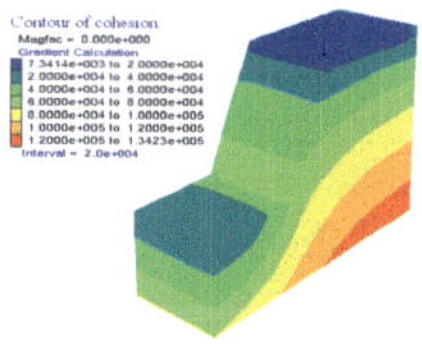	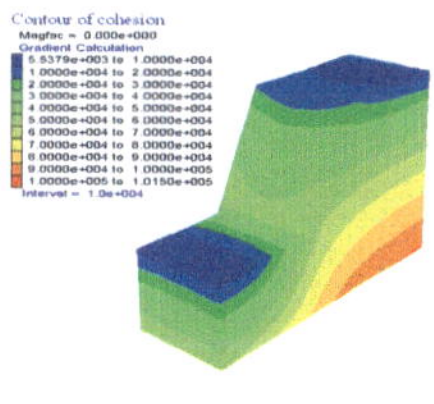	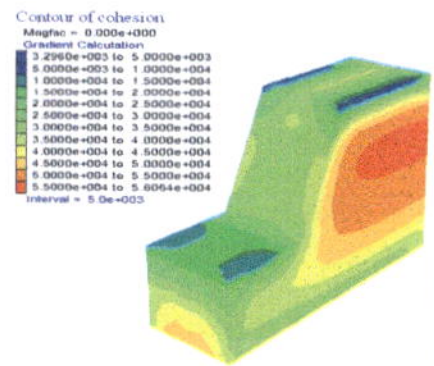
Contour of angle of friction	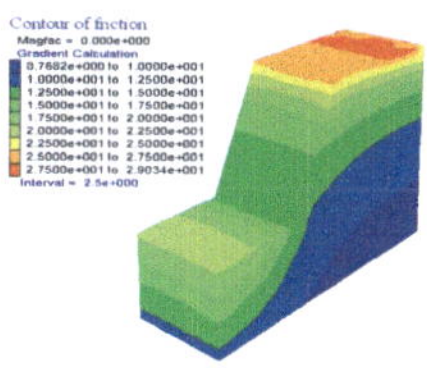	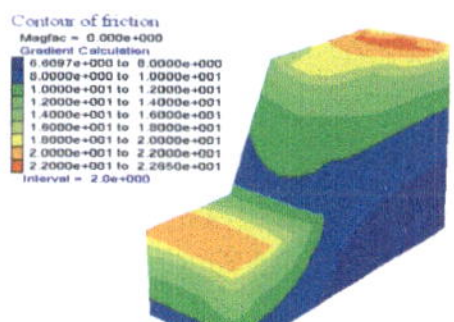	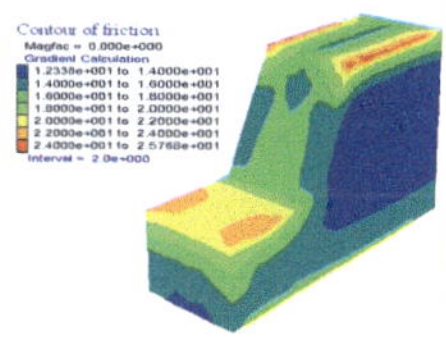
FOS	1.883	2.502	2.057
f_B		$f_{B,xy}$=1.329	$f_{B,xyz}$=1.092

Fig. 10.7 Comparison of failure surfaces, contours of c and ϕ and FOS of a slope model under various boundary conditions

FOS values, as well as the contours of instantaneous cohesion and angle of friction in a given slope under different boundary conditions.

The contours of the instantaneous c and ϕ- shown in Fig. 10.7 are calculated using Eqs. 10.5–10.8 together with the final stress states of each element when the slope failures, which can be used to illustrate the failure mechanics performed by the non-linear SSR technique. For boundary conditions 1 and 2, where the z-direction displacement of the end faces is not fixed, the slope surfaces have relatively higher ϕ values and lower c values compared with the values at the bottom of the slope. This disparity is a result of the fact that the stress state of the elements at the bottom of the slope is greater than the stress state of the elements near the slope surface; and the values of instantaneous c increase and ϕ decrease with the increase of σ_3 values as shown in Fig. 10.2.

For boundary condition 3, where x, y, z-direction displacement of the end faces is fully fixed, the contours of c and ϕ are obviously different from those of conditions 1 and 2. This can be explained by the fact that the stress state in the slope

Table 10.3 The results of FOS and fB of the slope with different slope angle

Slope angle (°)	Boundary condition used at end faces of slope model			Boundary weighting factor, f_B	
	FOS_y	FOS_{xy}	FOS_{xyz}	$f_{B,xy}$	$f_{B,xyz}$
30	2.11	3.75	3.42	1.78	1.62
45	2.06	2.97	2.52	1.44	1.22
60	1.88	2.50	2.06	1.33	1.09
75	1.57	2.11	1.73	1.34	1.10
90	1.29	1.69	1.43	1.31	1.11

under boundary condition 3 is different from the stress state in the slope under boundary conditions 1 and 2. Therefore, the values of instantaneous c and ϕ will change, which leads to the change of the shape of the failure surface, as well as the FOS values.

The value of the FOS for boundary condition 1 is equal to 1.883, which is lower than the FOS for boundary conditions 2 and 3 where the FOS is equal to 2.502 and 2.057, respectively. In order to investigate the possible correlation of the FOS under different boundary conditions, the authors proposed a boundary weighting factor, f_B, as shown in Fig. 10.7, which represents the ratio of the FOS from boundary conditions 2 and 3 to boundary condition 1. In this case, $f_{B,xy} = 2.502/1.883 = 1.329$ and $f_{B,xyz} = 2.057/1.883 = 1.092$.

Table 10.3 compares the FOS values under different boundary conditions for the slope (see Table 10.2), with the slope angle β varying from 30° to 90°. The correlations between f_B and β in Table 10.3 are plotted in Fig. 10.8. The figure demonstrates that the boundary weighting factor f_B decreases as the slope angle increases when $\beta < 50°$. However, f_B tends to reach stable values ($f_{B,xy} = 1.4$ and $f_{B,xyz} = 1.1$) when $\beta > 50°$.

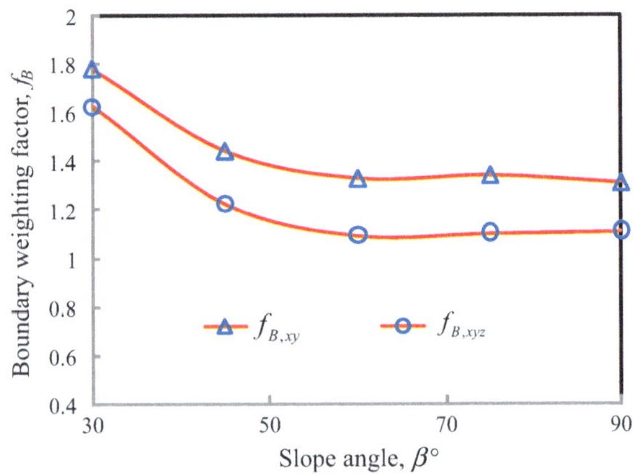

Fig. 10.8 The correlations between f_B and β under different boundary conditions for a slope case

Table 10.4 The results of FOS and fB of real slope cases

Cases	H (m)	β (°)	γ (kN/m^3)	σ_{ci} (MPa)	GSI	m_i^a	D^b	FOS_y	FOS_{xy}	FOS_{xyz}	$f_{B,xy}$	$f_{B,xyz}$
1	184	55	27	153	47	9	0.9	2.67	3.67	3.04	1.38	1.14
2	140	34	26	50	28	8	0.7	1.52	2.52	2.27	1.66	1.49
3	220	45	27	65	44	17	0.8	2.48	3.54	2.97	1.43	1.20
4	135	65	27	172	58	9	0.9	4.16	5.70	4.92	1.37	1.18
5	70	50	27	29	41	7	0.8	1.83	2.59	2.19	1.42	1.20
6	110	45	26.5	50	25	10	0.7	1.60	2.28	1.96	1.43	1.23
7	270	45	27	109	39	18	0.9	2.21	3.18	2.68	1.44	1.21
8	170	55	30	104	48	7	0.7	2.63	3.63	3.01	1.38	1.15
9	60	60	27	65	44	13	1	2.53	3.44	2.80	1.36	1.11
10	35	67	27	109	28	12	1	1.93	2.61	2.14	1.35	1.11
11	63	35	27	109	28	12	1	2.04	3.15	2.78	1.55	1.37
12	70	49	27	3	49	24	1	1.20	1.72	1.46	1.44	1.22
13	58	50	27	5	55	22	1	1.80	2.53	2.12	1.41	1.18
14	60	48	27	5	54	22	1	1.75	2.48	2.11	1.42	1.20
15	60	52	27	5	56	22	1	1.83	2.50	2.09	1.37	1.14
16	40	71	27	50	33	14	1	1.70	2.29	1.87	1.35	1.10
17	110	50	27	50	25	14	1	1.05	1.48	1.26	1.40	1.20
18	41	50	27	3	46	24	1	1.34	1.90	1.62	1.42	1.21
19	41	55	27	3	49	24	1	1.42	1.96	1.64	1.38	1.16
20	46	55	27	3	50	24	1	1.40	1.93	1.62	1.38	1.16
21	57	49	27	3	48	24	1	1.26	1.81	1.55	1.44	1.23
22[c]	57	37	27	3	48	24	1	1.36	2.13	1.88	1.56	1.38
23[c]	57	40	27	3	48	24	1	1.33	2.03	1.79	1.52	1.34
24[c]	57	42	27	3	48	24	1	1.32	1.97	1.72	1.49	1.30

[a]The values of m_i are estimated based on the information of uniaxial compressive strength for general rock type (Shen 2013)

[b]By considering the excavation methods for cases 9 and10 were poor and the cases 11–21 were obtained from open pit mines (the excavation method was assumed to be production blasting), therefore, according to the guidelines by Hoek et al. (2002), D was assumed to be 1 for all cases

[c]Cases 22–24 are additional cases which have the same rock mass properties as case 21 except slope angle

Results presented in Fig. 10.8 and Table 10.3, however, provide only one example of specific rock properties and slope height. To further check the correlation between f_B and β, the authors conducted a comprehensive study using 21 real cases collected from Douglas (2002) and Taheri and Tani (2010), with various slope geometries and rock mass properties as indicated in Table 10.4.

Our analysis of the database showed that there is a strong correlation between f_B and β as shown in Fig. 10.9. Most of the data lie along the lines which have a trend of decreasing f_B with the increase of β when $\beta < 50°$. When $\beta > 50°$, f_B tends to attain constant values. The results that were presented demonstrate that the effects of boundary conditions on the values of the FOS are more obvious for a slope with

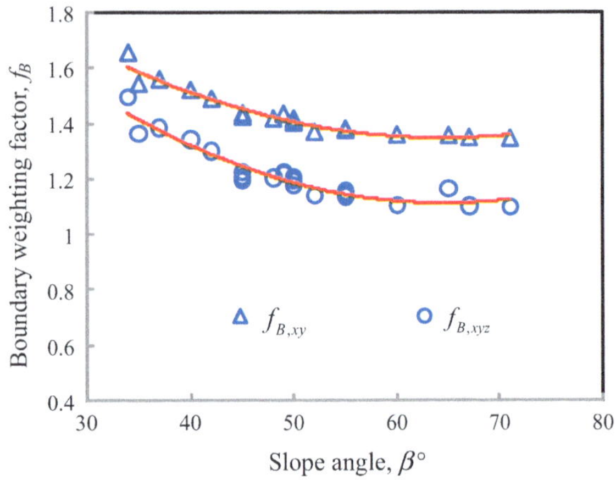

Fig. 10.9 The correlations between f_B and β under different boundary conditions for open pit cases

a low angle than a steep slope. The values of $f_{B,xy}$ and $f_{B,\ xyz}$ will go up to 1.7 and 1.5 when the slope angle is less than 35°. On the other hand, when the slope angle is more than 50°, the values of $f_{B,xy}$ and $f_{B,xyz}$ tend to equal 1.4 and 1.1, respectively.

The possible connections between f_B and other parameters (H, σ_{ci}, GSI and m_i) were also investigated as shown in Fig. 10.10. No strong relationship was observed between f_B and these parameters.

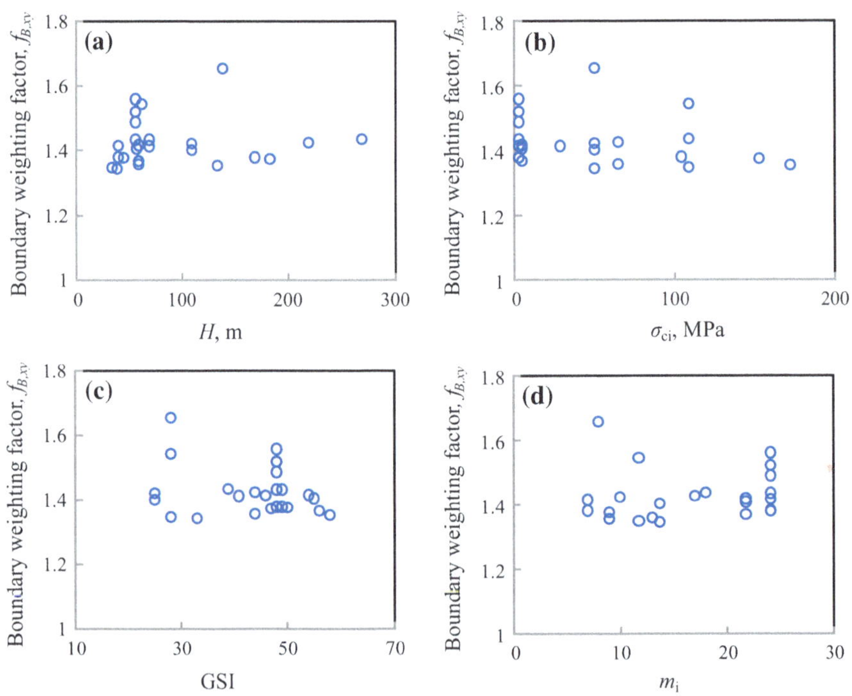

Fig. 10.10 The correlations between f_B,xy and H, σ_{ci}, GSI, m_i

References

Basarir H, Ozsan H, Karakus M (2005) Analysis of support requirements for a shallow diversion tunnel at Guledar dam site. Turkey Eng Geol 81(2):131–145

Cheng Y, Yip C (2007) Three-dimensional asymmetrical slope stability analysis extension of Bishop's, Janbu's, and Morgenstern-Price's techniques. J Geotech Geoenvironmental Eng 133 (12):1544–1555

Chugh A (2003) On the boundary conditions in slope stability analysis. Int J Numer Anal Meth Geomech 27:905–926

Dawson E, Roth W, Drescher A (1999) Slope stability analysis by strength reduction. Géotechnique 49(6):835–840

Detournay C, Hart R, Varona P (2011) Factor of safety measure for Hoek-Brown material. Continuum and distinct element numerical modeling in geomechanics-2011-Sainsbury, Hart, Detournay & Nelson (eds.) Paper: 13-07

Duncan J (1996) State of the art: limit equilibrium and finite-element analysis of slopes. J Geotech Geoenvironmental Eng 122(7):577–596

Douglas K (2002) The shear strength of rock masses. Ph.D. Thesis, University of New South Wales, Australia

Eid H (2010) Two-and three-dimensional analyses of translational slides in soils with nonlinear failure envelopes. Can Geotech J 47(4):388–399

Farzaneh O, Askari F, Ganjian N (2008) Three-dimensional stability analysis of convex slopes in plan view. J Geotech Geoenvironmental Eng 134(8):1192–1200

Fu W, Liao Y (2010) Non-linear shear strength reduction technique in slope stability calculation. Comput Geotech 37:288–298

Gharti H, Komatitsch D, Oye V, Martin R, Tromp J (2012) Application of an elastoplastic spectral-element method to 3D slope stability analysis. Int J Numer Meth Eng 91(1):1–26

Griffiths D, Marquez R (2007) Three-dimensional slope stability analysis by elasto-plastic finite elements. Géotechnique 57(6):537–546

Hammah R, Yacoub T, Corkum B, Curran J (2005) The shear strength reduction method for the generalized Hoek-Brown criterion. ARMA/USRMS 05-810

Hoek E, Brown ET (1980) Underground Excavations in Rock. Instn Min Metall, London

Hoek E, Carranza-Torres C, Corkum B (2002) Hoek-Brown failure criterion—2002 Edition. In: Hammah R, Bawden W, Curran J, Telesnicki M (eds), Proceedings of NARMS-TAC 2002, Mining innovation and technology. Toronto available from www.rocscience.com. 20 Sept 2011

Itasca Consulting Group (2009). FLAC3D 4.0 manual. Minneapolis

Jimenez R, Serrano A, Olalla C (2008) Linearization of the Hoek and Brown rock failure criterion for tunnelling in elasto-plastic rock masses. Int J Rock Mech Min Sci 45:1153–1163

Karakus M (2007) Appraising the methods accounting for 3D tunnelling effects in 2D plane strain FE analysis. Tunn Undergr Space Technol 22(1):47–56

Karakus M, Ozsan A, Basarir H (2007) Finite element analysis for the twin metro tunnel constructed in Ankara Clay, Turkey. Bull Eng Geol Environ 66(1):71–79

Kumar P (1998) Shear failure envelope of Hoek-Brown criterion for rockmass. Tunn Undergr Space Technol 13(4):453–458

Li A, Merifield R, Lyamin A (2008) Stability charts for rock slopes based on the Hoek-Brown failure criterion. Int J Rock Mech Min Sci 45(5):689–700

Li A, Merifield R, Lyamin A (2009) Limit analysis solutions for three dimensional undrained slopes. Comput Geotech 36(8):1330–1351

Li A, Merifield R, Lyamin A (2010) Three-dimensional stability charts for slopes based on limit analysis methods. Can Geotech J 47(12):1316–1334

Michalowski R (2010) Limit analysis and stability charts for 3D slope failures. J Geotech Geoenvironmental Eng 136(4):583–593

Michalowski R, Drescher A (2009) Three-dimensional stability of slopes and excavations. Géotechnique 59(10):839–850
Michalowski R, Nadukuru S (2013) Three-dimensional limit analysis of slopes with pore pressure. J Geotech Geoenvironmental Eng 139(9):1604–1610
Nadukuru S, Michalowski R (2013) Three-dimensional displacement analysis of slopes subjected to seismic loads. Can Geotech J 50(6):650–661
Naghadehi M, Jimenez R, Khalokakaie R, Jalali S (2013) A new open-pit mine slope instability index defined using the improved rock engineering systems approach. Int J Rock Mech Min Sci 61:1–14
Nian T, Huang R, Wan S, Chen G (2012) Three-dimensional strength-reduction finite element analysis of slopes: geometric effects. Can Geotech J 49(5):574–588
Priest S (2005) Determination of shear strength and three-dimensional yield strength for the Hoek-Brown criterion. Rock Mech Rock Eng 38(4):299–327
Stianson J, Fredlund D, Chan D (2011) Three-dimensional slope stability based on stresses from a stress-deformation analysis. Can Geotech J 48(6):891–904
Shen J (2013) Analytical and numerical analyses for rock slope stability using generalized Hoek-Brown criterion. Ph.D thesis, School of civil, Environmental and mining engineering, The University of Adelaide, Australia
Shen H, Abbas S (2013) Rock slope reliability analysis based on distinct element method and random set theory. Int J Rock Mech Min Sci 61:15–22
Shen J, Karakus M, Xu (2012a) Direct expressions for linearization of shear strength envelopes given by the generalized Hoek-Brown criterion using genetic programming. Comput Geotech 44:139–146
Shen J, Priest S, Karakus M (2012b) Determination of Mohr-Coulomb shear strength parameters from generalized Hoek-Brown criterion for slope stability analysis. Rock Mech Rock Eng 45:123–129
Taheri A, Tani K (2010) Assessment of the stability of rock slopes by the slope stability rating classification system. Rock Mech Rock Eng 43:321–333
Tao M, Li X, Li D (2013) Rock failure induced by dynamic unloading under 3D stress state. Theoret Appl Fract Mech 65:47–54
Tutluoglu L, Ferid I, Karpuz C (2011) Two and three dimensional analysis of a slope failure in a lignite mine. Comput Geosci 37(2):232–240
Wei W, Cheng Y, Li L (2009) Three-dimensional slope failure analysis by the strength reduction and limit equilibrium methods. Comput Geotech 36(1–2):70–80
Zhang Y, Chen G, Zheng L, Li Y, Zhuang X (2013) Effects of geometries on three-dimensional slope stability. Can Geotech J 50(3):223–249
Zheng H (2012) A three-dimensional rigorous method for stability analysis of landslides. Eng Geol 145–146:30–40

Chapter 11
Chart-Based Slope Stability Assessment Using the Generalized Hoek–Brown Criterion

Abstract Slope stability charts are used extensively in practical application to meet the need of quick assessment of rock slope design. However, rock slope stability charts based on the Generalized Hoek–Brown (GHB) criterion, which is one of the most widely adopted failure criteria to estimate rock mass strength in rock engineering, are considerably limited. This paper presents new stability charts for the analysis of rock mass slopes satisfying with the GHB criterion. Firstly, charts for calculating the factor of safety (FOS) of a slope for a specified slope angle $\beta = 45°$ are proposed. Secondly, a disturbance weighting factor f_D is introduced to illustrate the effect of disturbance factor D upon the stability of rock slopes. Thirdly, a slope angle weighting factor f_β is proposed to show the influence of slope angle β on slope stability. Combined with stability charts based on $\beta = 45°$, the weighting factors f_D and f_β allow the calculation of the FOS of a slope assigned various slope angle under different blasting damage and stress relief conditions. The reliability of the proposed charts is tested against numerical solutions. The results show that FOS from the proposed charts exhibits only 3.1% average discrepancy from numerical solutions using 1680 sets of data. The proposed charts are simple and straightforward to use and can be adopted as useful tools for the preliminary rock slope stability analysis.

11.1 Introduction

Determining the stability of rock mass slopes is an important task in many areas of civil and mining engineering, such as open pit excavation and large dam construction. Most slope stability analysis is based on seeking the factor of safety (FOS), which is a traditional measure of the safety margin of a given slope (Michalowski 2010). Having the merit of quick assessment of preliminary slope design, stability charts have been extensively used to estimate the stability of a slope in practical applications. The most common charts widely used in slope engineering is the Taylor's stability charts (Taylor 1937), which require the Mohr–Coulomb (MC) shear strength parameters cohesion c and angle of friction ϕ to

J. Zuo and J. Shen, *The Hoek-Brown Failure criterion—From theory to application*,
https://doi.org/10.1007/978-981-15-1769-3_11

estimate the FOS of a slope. However, rock mass strength is a non-linear stress function, therefore, the linear MC criterion generally do not agree with the rock mass failure envelope (Jimenez et al. 2008; Sheorey 1997; Shen et al. 2012a; Hoek and Brown 1980), especially for slope stability problems where the rock mass is in a state of low confining stresses that make the nonlinearity more obvious.

Currently, the Hoek–Brown (HB) criterion (Hoek and Brown 1997) is one of the most broadly adopted failure criteria to estimate rock mass strength in rock engineering. Over the past 30 years the HB criterion has been applied successfully to a wide range of intact and fractured rock types. The latest version is the Generalized Hoek–Brown (GHB) criterion presented by Hoek et al. (2002). The equations are expressed as follows:

$$\sigma_1 = \sigma_3 + \sigma_{ci}\left(\frac{m_b\sigma_3}{\sigma_{ci}} + s\right)^a \tag{11.1}$$

m_b, s and a are the Hoek–Brown input parameters that depend on the degree of fracturing of the rock mass and can be estimated from the Geological Strength Index (GSI), given by:

$$m_b = m_i e^{\left(\frac{GSI-100}{28-14D}\right)} \tag{11.2}$$

$$s = e^{\left(\frac{GSI-100}{9-3D}\right)} \tag{11.3}$$

$$a = 0.5 + \frac{e^{\left(\frac{-GSI}{15}\right)} - e^{\left(\frac{-20}{3}\right)}}{6} \tag{11.4}$$

where, σ_1 is the maximum principal stresses, σ_3 is the minimum principal stresses, σ_{ci} is the uniaxial compressive strength of the intact rock, m_{i} is the Hoek–Brown constant of the intact rock, and D is the disturbance factor of the rock mass. The input parameters of the GHB criterion can be achieved directly from mineralogical assessment, uniaxial compressive testing of rock materials, and measurement of discontinuities characteristic of rock masses (Priest 2005). Therefore, a great advance in the field of rock slope stability assessment could be achieved if suitable stability charts could be developed to estimate the FOS directly from the GHB criterion.

Development of rock slope stability charts based on the GHB criterion, however, is a challenging task since there are at least six input parameters (GSI, m_{i}, σ_{ci}, γ, β, H) involved to calculate the FOS for a given dry slope with $D = 0$, where, γ is the unit weight of the rock mass, β is the slope angle, and H is the slope height. Based on our literature review, charts for the estimation of FOS directly from the GHB criterion is still a under research area and very few charts are available in the literature.

In the Chapter, we present new charts which can be used to estimate the FOS of a slope directly from the Hoek–Brown parameters (GSI, m_{i} and D), slope geometry

(β and H) and rock mass properties (σ_{ci} and γ). The proposed charts are straightforward to use and can be adopted as useful tools for the preliminary rock slope stability assessment.

In this Chapter, the existing rock slope stability charts related to the HB criterion are briefly reviewed in Sect. 11.2. The proposed stability charts are presented in Sect. 11.3. Charts application to slope cases is presented in Sect. 11.4.

11.2 Review of Existing Rock Slope Stability Charts Based on the HB Criterion

Since Taylor (1937) proposed a set of stability charts for soil slopes, chart solutions have been presented by many researchers (Baker 2003; Hoek and Bray 1981; Carranza-Torres 2004; Li et al. 2008, 2009, 2011; Zanbak 1983; Naghadehi et al. 2013; Steward et al. 2011; Wyllie and Mah 2004; Taheri and Tani 2010; Baker et al. 2006) and are still widely used as design tools in slope engineering. At present, rock slope stability charts, such as Hoek and Bray (1981)'s, often need to use the equivalent MC shear strength parameters cohesion c and angle of friction ϕ, which can be estimated from software *RocData* as shown in Fig. 11.1. The equivalent fitting MC envelope is a straight line. The slope of the tangent to the MC envelope gives the value of ϕ, and the intercept with the shear stress axis gives the value of c. However, this conversion has been found to yield inconsistent estimates of the FOS of a slope, with a discrepancy between the HB and equivalent MC envelopes of up to 64% (Li et al. 2008).

Until now, the slope stability charts by Carranza-Torres (2004) and Li et al. (2008, 2009, 2011) have been among the few charts that can be used to estimate the FOS directly from the HB criterion. Carranza-Torres (2004) proposed a solution for

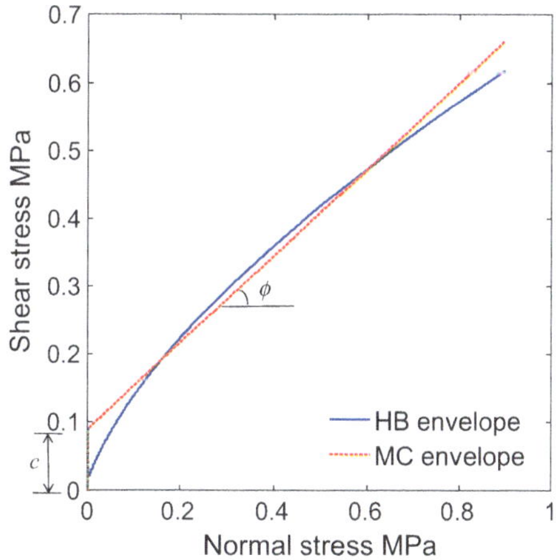

Fig. 11.1 Relationship between HB and equivalent MC envelopes

estimating the shear strength of rock masses from the HB criterion, which was incorporated in the Bishop simplified method (Shen et al. 2012a) for the analysis of rock slope stability. Carranza-Torres (2004) revealed that when the Hoek–Brown parameter $a = 0.5$, the FOS of a given slope only depends on the three independent variables, $\overline{\gamma H}$, s/m_b^2 and β.

$$\overline{\gamma H} = \frac{\gamma H}{\sigma_{ci} m_b} + \frac{s}{m_b^2} \quad (11.5)$$

In order to estimate the FOS of a slope with a given geometry (β and H), rock mass properties (γ and σ_{ci}) and Hoek–Brown parameters (GSI, m_i and D), firstly, the values of m_b and s can be calculated using Eqs. (11.2) and (11.3), respectively. Thereafter, the values of the dimensionless parameters $\overline{\gamma H}$ and s/m_b^2 can be calculated. Finally, the FOS can be directly estimated from the chart as shown in Fig. 11.2.

While Carranza-Torres (2004) proposed a chart is based on $a = 0.5$ with only a single slope angle, $\beta = 45°$, the current research proposed a slope angle weighting factor f_β to illustrate the influence of slope angle β on slope stability, to be discussed in Sect. 11.3. Combined with the Fig. 11.2 based on $\beta = 45°$, the slope angle weighting factor chart (as shown in Fig. 11.11) can be used for estimating the FOS of a slope assigned various angles.

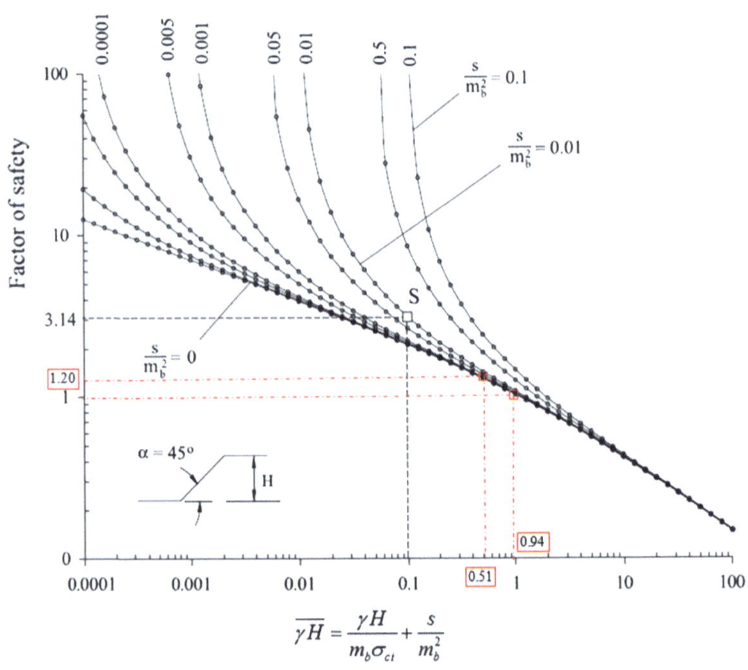

Fig. 11.2 Slope stability chart ($\beta = 45°$, $a = 0.5$) (Jimenez et al. 2008)

Stability charts for estimating rock mass slopes directly from the Hoek–Brown parameters GSI, mi and D were originally proposed by Li et al. (2008) using limit analysis (LA). Li et al.'s charts are based on the assumption D = 0, which means that the rock slope is undisturbed. Similar stability charts utilizing D = 0.7 and D = 1.0 were also proposed by Li et al. (2011) in order to examine the effects of these rates of disturbance on rock slope stability. Seismic stability charts were also proposed by Li et al. (2009) to account for seismic effects on rock slope stability. The current research, however, focuses on static slope stability analysis, and seismic charts were not discussed in detail here.

Figure 11.3 shows typical stability charts with a slope angle of $\beta = 45°$. N is the non-dimensional stability number, defined as:

$$N = \frac{\sigma_{ci}}{\gamma H FOS_{LA}} \tag{11.6}$$

Because the upper and lower boundary results bracket a narrow range of N within ±9% or better, Li et al. (2009, 2011) adopted the average value limit solution to generate the charts in order to keep their calculations simple. The use of these charts is quite easy. Firstly, the stability number N can be calculated using the values of GSI and m_i from an appropriate chart according to a specified D value ($D = 0$, 0.7 and 1) as shown in Fig. 11.3. Having obtained the value of N, Eq. (11.6) can be used to calculate the FOS_{LA}.

As noted by Li et al. (2012), the definition of factor of safety for Eq. (11.6) is different from that of FOS obtained from the limit equilibrium method (LEM) which was defined as a function of resisting force f_R divided by driving force f_D, $FOS_{LEM} = f_R/f_D$. Therefore, the values of FOS_{LA} obtained from Eq. (11.6) are generally not equal to FOS_{LEM}. Such variations can be illustrated using four examples in Table 11.1.

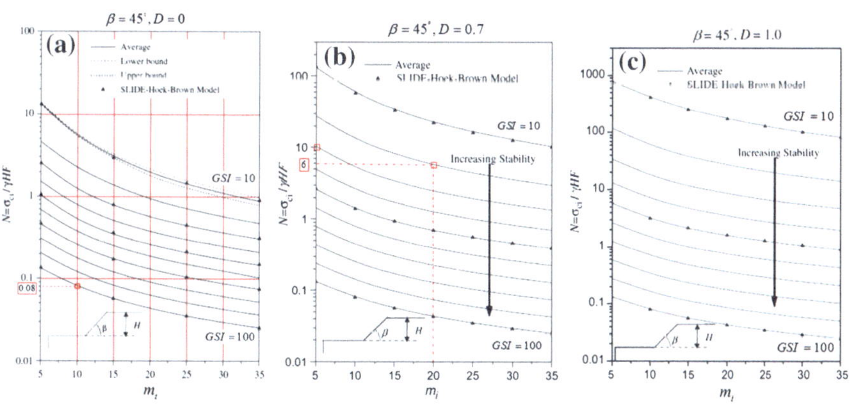

Fig. 11.3 **a** Slope stability chart with $D = 0.7$ (Li et al. 2009); **b** slope stability chart with D = 0.7 (Li et al. 2011); **c** slope stability chart with D = 1.0 (Li et al. 2011)

Table 11.1 Comparison of the factor of safety estimated from different stability charts

Input parameters	Example 1 (Carranza-Torres 2004)	Example 2 (Li et al. 2011)	Example 3	Example 4
σ_{ci}, MPa	0.75	10	13.5	5.4
GSI	100	30	30	20
m_i	10	8	5	20
D	0	1	0.7	0.7
γ, kN/m^3	25	23	27	27
H, m	27	50	50	25
β, °	45	60	45	45
Calculated parameters				
m_b	10	0.054	0.107	0.247
s	1	8.57E–06	3.93E–05	9.22E–06
a	0.5	0.522	0.522	0.544
s/m_b^2	0.0100	0.0030	0.0034	0.0002
$\overline{\gamma H}$	0.10	2.13	0.94	0.51
$\sigma_{ci}/(\gamma H)$	1.11	8.70	10	8
N	0.08	50	10	6
Factor of safety				
Carranza-Torres Chart (LEM)	3.14	0.44[a]	1.00	1.20
Li et al. Chart (LA)	13.89	0.17	1.00	1.33

[a]FOS of Example 2 is calculated from Fig. 11.2 together with Fig. 11.11

Considering that the LEM is still the most widely used methods for slope stability analysis, we proposed an alternative stability chart based on the LEM. The proposed charts are able to estimate the FOS of a slope directly from the Hoek–Brown parameters (GSI, m_i and D), slope geometry (β and H) and rock mass properties (σ_{ci} and γ).

11.3 Proposed Stability Charts for Rock Mass Slopes

The work outlined here required hundreds of runs on a microcomputer, analyzing the stability of various slopes having different geometries and rock mass properties. The slope models were analyzed using *Slide 6.0*. Details of the slope model settings are shown in Table 11.2.

The competency factor, the ratio of the uniaxial compressive strength σ_{ci} to the pressure of the overburden γH of tunnels, proposed by Muirwood (1972) was used in current study. For rock slope application, γH can represent the vertical stress of the rock slope, in this paper, $\sigma_{ci}/(\gamma H)$ was termed the strength ratio (SR) of a rock slope. We interested the stability number N proposed by Li et al. (2008), which

Table 11.2 Slope modeling setting in *Slide 6.0*

Modeling setting summary	
Analysis method	Bishop simplified
Number of slides	25
Search method	Auto grid search
Rock strength type	Generalized Hoek–Brown
Ground water	None
Failure surface type	Circular toe failure
Disturbance factor	0–1

contains SR as shown Eq. (11.6). The use of SR is a significant innovation for the rock slope stability analysis because when the values of the input parameters GSI, m_i, D and β are determined, the FOS is related only to the SR of that slope.

In this section, the derivation of the theoretical relationship between the SR and FOS of a given slope slip surface will be explained in detail. In the next stage, based on the relationship between the SR and FOS, charts for calculating the FOS_{45° when $\beta = 45°$, $D = 0$ are proposed. A disturbance weighting factor f_D is then introduced to illustrate the effect of disturbance D upon the stability of rock mass slopes. Finally, a slope angle weighting factor f_β is proposed to illustrate the influence of the slope angle β on slope stability. Combined with stability charts based on $\beta = 45°$, the weighting factors f_β and f_D allow for the calculation of the FOS of slopes exhibiting various angle under different blast damage and stress relief conditions. Also, some slope examples are presented to illustrate the use of the proposed charts.

11.3.1 Theoretical Relationship Between SR and FOS

Combined with a generic form of Balmer's equations (Carranza-Torres 2004), the GHB criterion was input into *Slide 6.0* in order to calculate the instantaneous shear stress τ of each slice of a failure surface under a specified normal stress σ_n. The generic form of Balmer's equations are expressed as follows:

$$\frac{\sigma_n}{\sigma_{ci}} = \frac{\sigma_3}{\sigma_{ci}} + \frac{1}{2}\left(m_b\frac{\sigma_3}{\sigma_{ci}} + s\right)^a\left[1 - \frac{am_b\left(m_b\frac{\sigma_3}{\sigma_{ci}} + s\right)^{a-1}}{2 + am_b\left(m_b\frac{\sigma_3}{\sigma_{ci}} + s\right)^{a-1}}\right] \tag{11.7}$$

$$\frac{\tau}{\sigma_{ci}} = \left(m_b\frac{\sigma_3}{\sigma_{ci}} + s\right)^a\frac{\sqrt{1 + am_b\left(m_b\frac{\sigma_3}{\sigma_{ci}} + s\right)^{a-1}}}{2 + am_b\left(m_b\frac{\sigma_3}{\sigma_{ci}} + s\right)^{a-1}} \tag{11.8}$$

For a generic case $a \neq 0.5$, in order to calculate shear stress τ, for the given values of the input parameters m_b, s, a, σ_{ci} and σ_n, Eq. (11.7) is solved iteratively to calculate the σ_3 value (Shen et al. 2012b). Having obtained σ_3, Eq. (11.8) can be used to calculate shear stress τ, therefore, σ_3/σ_{ci} can be expressed as follows:

$$\frac{\sigma_3}{\sigma_{ci}} = f_1\left(\frac{\sigma_n}{\sigma_{ci}}, m_b, s, a\right) \tag{11.9}$$

Also, τ/σ_{ci} can be expressed as follows:

$$\frac{\tau}{\sigma_{ci}} = f_2\left(\frac{\sigma_n}{\sigma_{ci}}, m_b, s, a\right) \tag{11.10}$$

The FOS can be defined as a function of resisting force f_R divided by driving force f_D as shown in Fig. 11.4a. The forces of f_R and f_D can also be expressed in terms of τ and σ_n acting on the base of an arbitrary slice i as shown in Fig. 11.4b. Resisting shear stress $\tau_R{}^i$ of the rock mass is governed by Eq. (11.10), and driving shear stress $\tau_D{}^i$ will depend on the weight of the slice γh^i as indicated in Fig. 11.4b.

Therefore, with the help of Eq. (11.10), the FOS can be expressed as follows:

$$FOS = f_3\left(\frac{\tau}{\gamma h}\right) = f_3\left[\frac{\sigma_{ci}}{\gamma h} f_2\left(\frac{\sigma_n}{\sigma_{ci}}, m_b, s, a\right)\right] \tag{11.11}$$

The value of σ_n of arbitrary slice depends on the weight of slice γh, which in turn depends on the characteristic stress γH and slope angle β (Carranza-Torres 2004). Equation (11.11), therefore, can be transformed into Eq. (11.12).

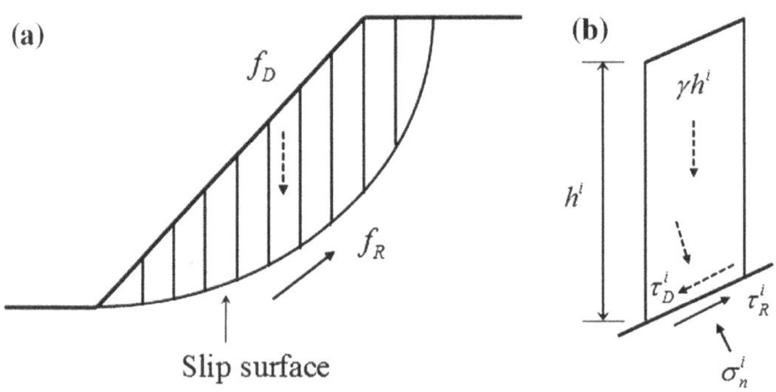

Fig. 11.4 **a** The basic of method of slices; **b** stresses acting on a given slice

$$FOS = f_5\left[\frac{\sigma_{ci}}{\gamma H} f_4\left(\frac{\gamma H}{\sigma_{ci}}, m_b, s, a, \beta\right)\right] \tag{11.12}$$

The parameters m_b, s and a in Eq. (11.12) can be calculated from Eqs. (11.2) to (11.4), respectively. Finally, the FOS can be expressed as Eq. (11.13)

$$FOS = f_6\left(\frac{\sigma_{ci}}{\gamma H}, GSI, m_i, D, \beta\right) = f_6(SR, GSI, m_i, D, \beta) \tag{11.13}$$

Equation (11.13) illustrates the fact that when the values of GSI, m_i and D are given in a homogeneous slope, along with the slope angle β, the FOS of a slip surface for a particular method of slices is uniquely related to the dimensionless parameter SR regardless of the magnitude of individual parameters σ_{ci}, γ and H.

Table 11.3 shows three different groups of σ_{ci}, γ and H associated with the same SR value for a slope that has the same values of GSI, m_i, D and β. The values of FOS were calculated using four limit equilibrium methods in *Slide 6.0,* with finite element method (FEM) conducted using the program *Phase*^{2} *8.0.*

The results show that FOS values for all three groups are exactly the same. Results of the comparison of the FOS calculated for the three groups over a range of GSI and m_i are shown in Table 11.4. Again, the results reveal that the FOS of a slope depends only on the magnitude of SR when the values of β, GSI, m_i and D are the same. Based on the relationship between the SR and FOS, the number of independent parameters for calculating the FOS can be reduced to four (SR, GSI, m_i and β) when $D = 0$. In the next stage, we will propose the rock slope stability charts based on the SR, GSI and m_i for slopes with $\beta = 45°$, $D = 0$.

Table 11.3 Comparison of the FOS of a given slope with the same value of SR

Input parameters	Group 1 (Li et al. 2008)	Group 2	Group 3
GSI	30	30	30
m_i	8	8	8
D	0	0	0
β, °	60	60	60
σ_{ci}, MPa	20	25	250
γ, kN/m^3	23	28.75	23.96
H, m	25	25	300
$\sigma_{ci}/(\gamma H)$	34.783	34.783	34.783
Factor of safety			
Bishop simplified	2.026	2.026	2.026
Janbu simplified	1.934	1.934	1.934
Spencer	2.032	2.032	2.032
Morgenstern-Price	2.027	2.027	2.027
Phase^2 *8.0* (FEM)	2.000	2.040	2.030

Table 11.4 Comparison of the FOS of a given rock slope with various Hoek–Brown parameters

Hoek–Brown parameters		Group 1	Group 2	Group 3
GSI	m_i	FOS	FOS	FOS
10	5	0.958	0.958	0.958
10	15	1.326	1.326	1.326
10	25	1.547	1.547	1.548
10	35	1.705	1.705	1.706
40	5	2.532	2.532	2.532
40	15	2.819	2.819	2.819
40	25	3.043	3.043	3.043
40	35	3.227	3.227	3.227
100	5	46.854	46.854	46.856
100	15	30.840	30.840	30.842
100	25	25.540	25.540	25.542
100	35	22.753	22.753	22.755

11.3.2 Slope Stability Charts Based on Slope Angle β = 45°

The examination of 54 slope case histories (Taheri and Tani 2010) from Iran and Australia shows that the average slope angle is 46.3°. Therefore, firstly, the proposed stability charts for the current study were based on the GHB criterion from a range of SR, GSI and m_i, but with a specified slope angle $\beta = 45°$ and $D = 0$ as shown in Fig. 11.5.

Figure 11.5 indicates that there is a clear trend of the increase of FOS with the increase of GSI and SR. For example, increasing GSI values from 10 to 100 when SR = 1, the values of FOS increase from 0.45 to 2.80 as shown in Fig. 11.5a. It is also found that SR has a considerable effect on the FOS, especially, under the state of high GSI values. For example, when GSI = 90, the values of FOS are equal to 3.1 for SR = 2 and increase to 5.7 for SR = 5 in Fig. 11.5a. On the other hand, under the state of low GSI values, there is a moderate increase of FOS with the increase of SR. For instance, when GSI=10, the value of FOS is equal to 1 for SR = 15, and FOS increase to 1.4 for SR = 40 as shown in Fig. 11.5a.

Alternative form of stability charts are shown in Fig. 11.6. We can see that, overall, the FOS increase with the increase of m_i values. However, at the state of high GSI and SR values, the FOS decrease with the increase of m_i values as shown in Fig. 11.6c, d. This phenomenon can be explained by Fig. 11.7 (Cai et al. 2004), which illustrates the relationship between the Mohr-Coulomb shear strength parameters and Hoek–Brown parameters GSI and m_i values. It is clear that the values of cohesion decrease with the increase of m_i values when $60 < \text{GSI} < 90$. Therefore, the resisting shear strength will decrease, which leads to the decrease of FOS, when m_i values increase in these specified cases.

Figure 11.8 is an alternative form of Fig. 11.6a using the stability number N proposed by Li et al. (2008). It should be noted that the values of N obtained from

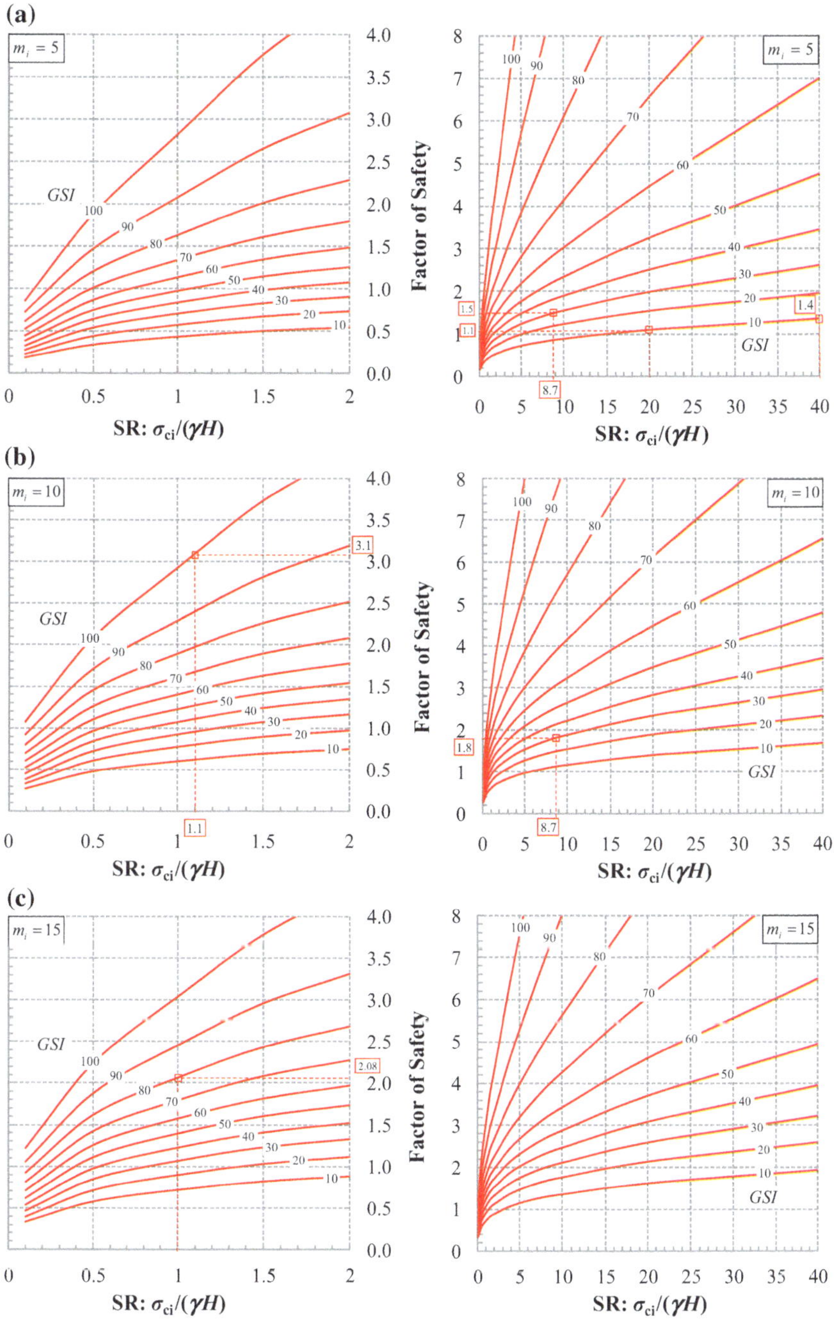

Fig. 11.5 Proposed stability charts for rock mass slope, $\beta = 45°$, $D = 0$ ($5 \le m_i \le 35$)

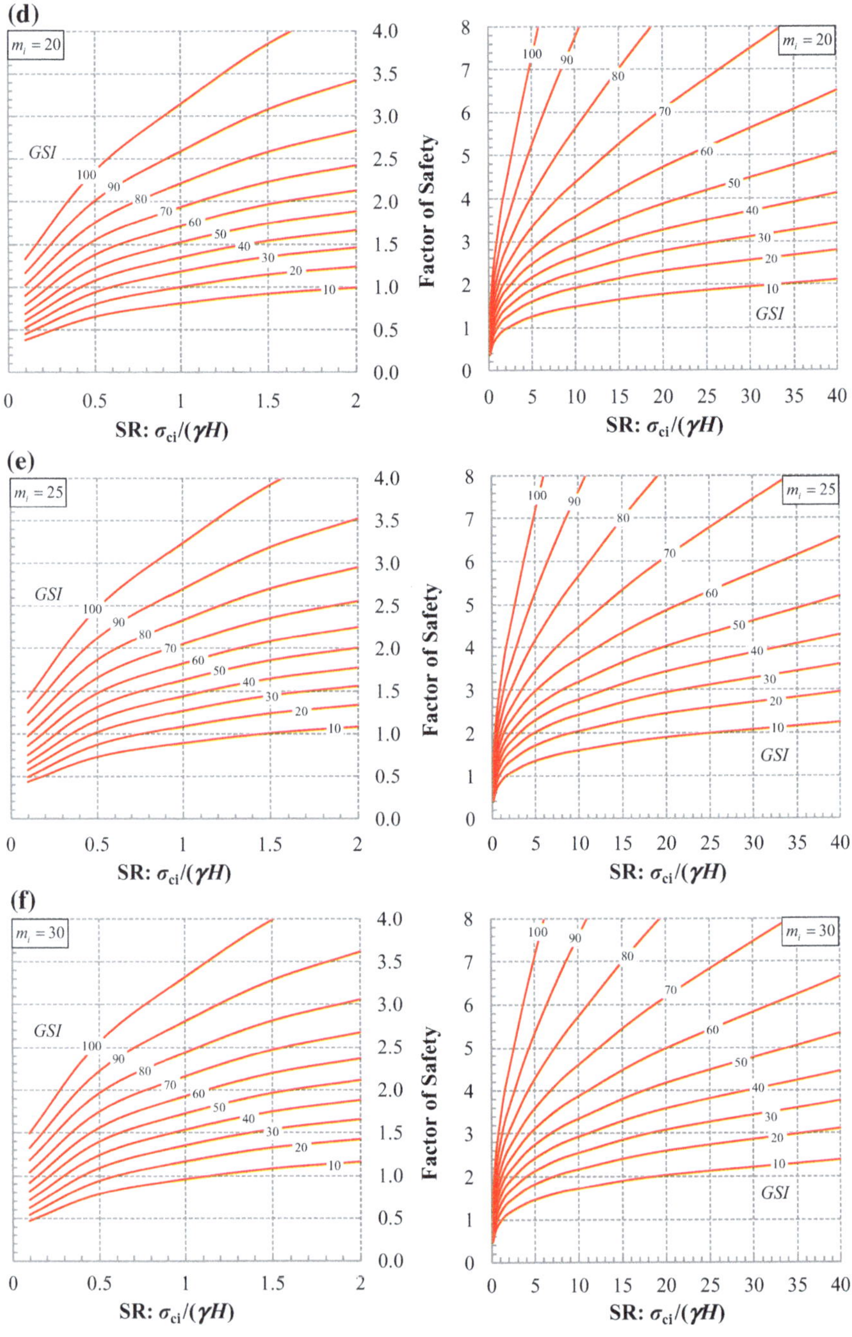

Fig. 11.5 (continued)

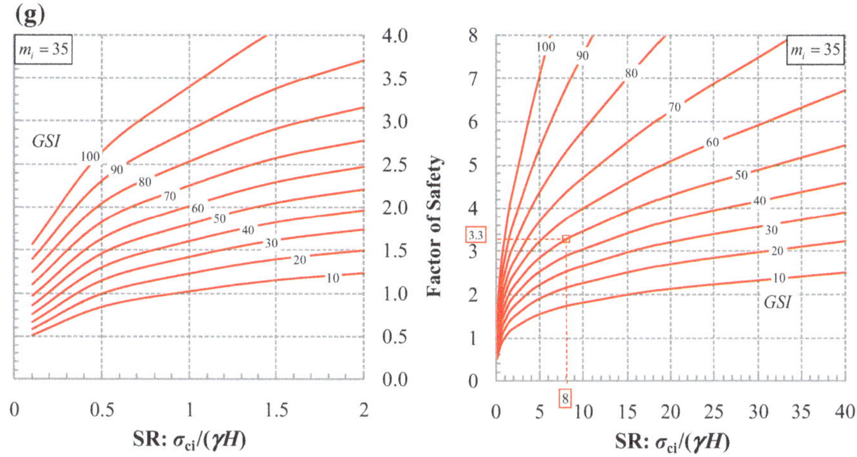

Fig. 11.5 (continued)

Fig. 11.8 are different from Fig. 11.3a, as the FOS calculated from limit equilibrium method are generally not equal to those from limit analysis (Li et al. 2012).

11.3.3 The Disturbance Weighting Factor f_D

Practical experience in the design of large open pit slopes has demonstrated that the estimation of rock mass properties from the HB criterion are too optimistic when $D = 0$ because of the realities of rock mass disturbance. Therefore, Hoek et al. (2002) introduced the disturbance factor D, which can vary from zero for undisturbed in situ rock masses to one for highly disturbed rock masses, to consider the effects of heavy blast damage as well as stress relief result in disturbance of the rock mass.

It is not easy to determine the exact value of D as various factors can influence the degree of disturbance in the rock mass. Hoek et al. (2002), Hoek and Diederichs (2006) and Hoek (2007), therefore, presented a number of slope cases to illustrate how to choose an appropriate D value for practical application. In civil engineering, small scale rock slope blasting results in modest rock mass damage, $D = 1.0$ and 0.7 were recommended under poor blasting and good blasting, respectively. For folded sedimentary rocks in a carefully excavated road cutting, $D = 0.3$ is suggested since the disturbance is relatively shallow. In mining engineering, large scale open pit mine slopes suffer significant disturbance under heavy production blasting, and $D = 1.0$ is the suggested value. For softer rocks under mechanical excavation, D is assumed to equal to 0.7. However, these guidelines are based on a limited number of case histories, and it can be argued that they should be extended and modified by considering more cases obtained from practical applications (Sonmez et al. 2006).

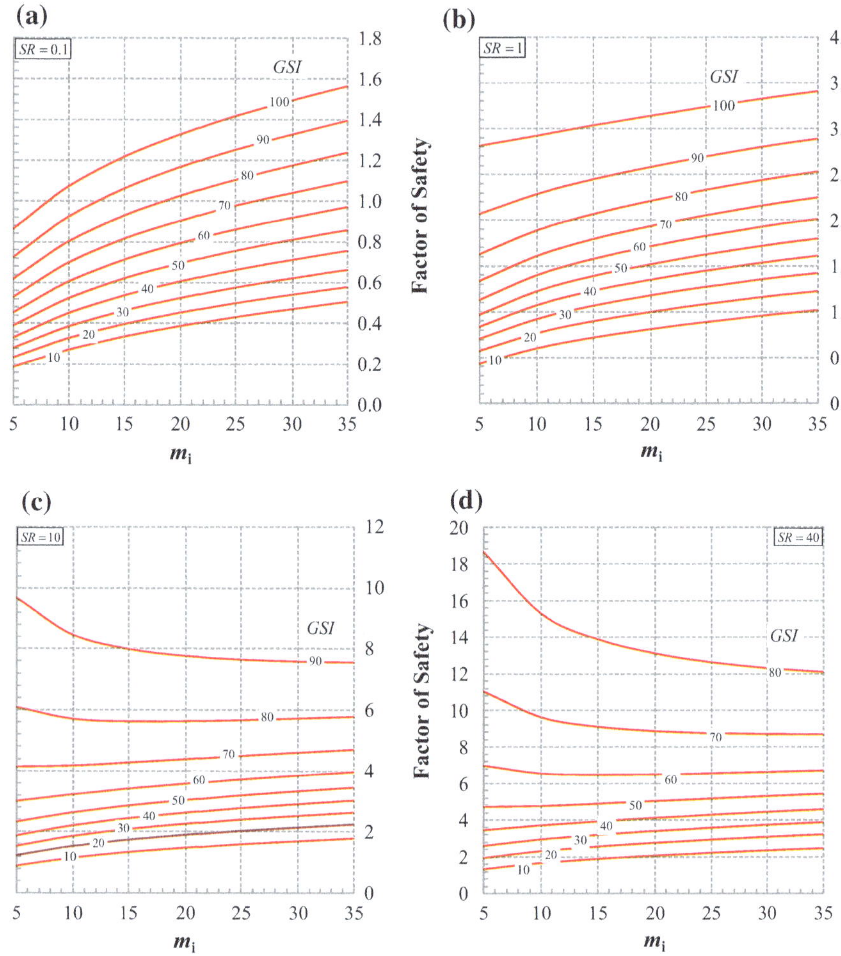

Fig. 11.6 Proposed stability charts for rock mass slope, $\beta = 45°$, $D = 0$ (SR = 0.1, 1, 10, 40)

Thus, in order to understand the real influence of D upon the stability of rock mass slopes, it is critical that researchers and engineers perform studies of a range of D values rather rely on the results from a single D analysis. As noted by Hoek and Diederichs (2006), the sensitivity analysis of a design is probably more significant in judging the acceptability of the design than a single calculated FOS.

The current study, therefore, proposes a disturbance weighting factor f_D to use in refining the influence of D upon the stability of rock mass slopes. The first step in proving the importance of factor f_D in determining the influence of D in calculating the FOS is to assume a disturbance factor D from 0 to 1, using the same values of GSI, m_i, SR and β as those found in the slope models in Sect. 11.3.2 with $D = 0$. Figure 11.9 illustrates the relationship between D and f_D for a slope with SR = 10

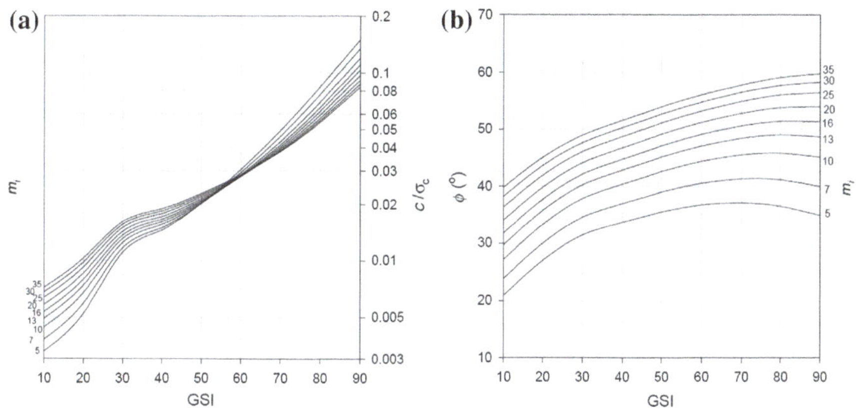

Fig. 11.7 **a** Relationship between c/σ_{ci} and GSI for different m_i values (Taylor 1937); **b** relationship between ϕ and GSI for different m_i values (Taylor 1937)

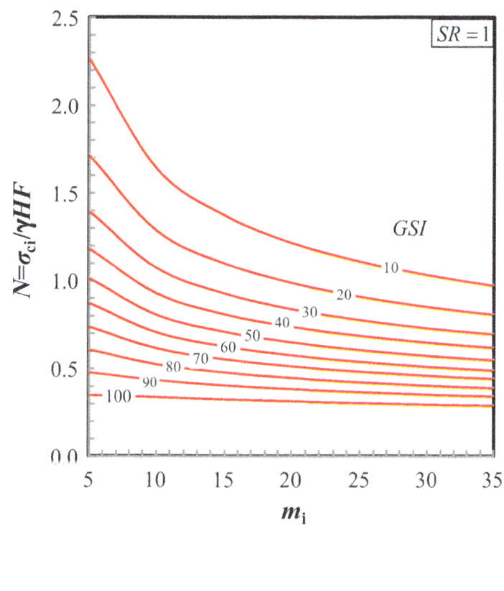

Fig. 11.8 Alternative form of Fig. 11.6b using the stability number N

and $\beta = 45°$. It is found that the minimum value of $m_i = 5$ and the maximum value of $m_i = 35$ generates a narrow range of f_D, which indicates that the value of m_i has an insignificant influence upon the estimation of f_D. For example, for a slope with GSI = 10, SR = 10, $\beta = 45°$, $D = 0.7$, increasing m_i from 5 to 35 only results in an increase in f_D from 0.42 to 0.48.

By considering the limit influence of m_i on the estimation of f_D, charts representing the relationship between f_D and D based on m_i = 5, 15, 25 and 35 were proposed, as shown in Fig. 11.10. The use of Fig. 11.10 to calculate the value of f_D is easy. For example, for a given slope with GSI = 90, $D = 0.7$, SR = 10 and $m_i = 5$, the value of f_D is equal to 0.88 as shown in Fig. 11.10a.

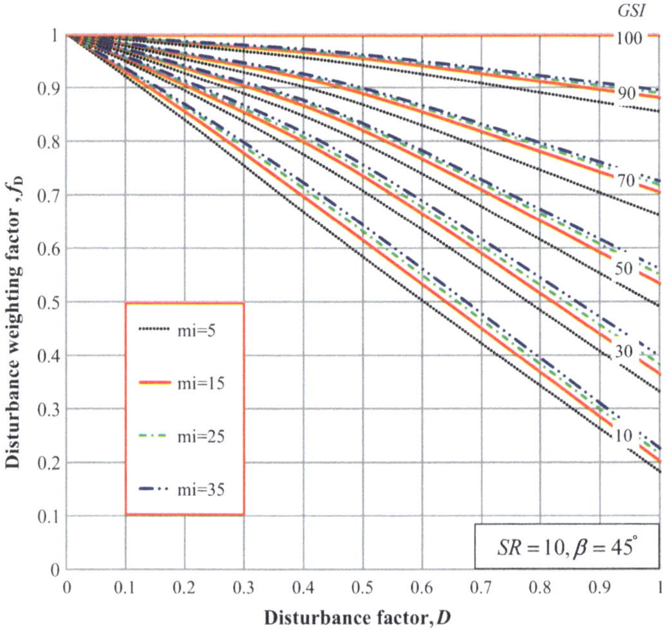

Fig. 11.9 Chart for estimating disturbance weighting factor f_D, SR = 10, β = 45°

11.3.4 *The Slope Angle Weighting Factor* $\mathbf{f}_\beta$

The values of FOS_{45° estimated directly from the data in Fig. 11.5 are based on a slope angle β = 45°. In order to examine the influence on the FOS of the slope angle, it was necessary to test the slope models using angles of different values. Slope angle β was assigned values ranging from 30° to 75° while the values of the GSI, m_i, D and SR are the same as slope models with β = 45°.

After hundreds of computer runs using a wide range of rock mass properties and slope geometry, a chart representing the relationship between the slope angle weighting factor f_β and the slope angle β was proposed based on the data $0 < \text{FOS} < 4$, which will be applicable for most civil and mining slope cases, as shown in Fig. 11.11.

By adopting a curve fitting strategy, a simplified equation was developed, as shown in Eq. (11.14).

$$f_\beta = 2.66e^{-0.022\beta} \tag{11.14}$$

Combined with the stability charts (Figs. 11.5 and 11.10), the slope angle weighting factor chart or Eq. (11.14) can be used for estimating the FOS of a slope with various given slope angle β estimated from real cases.

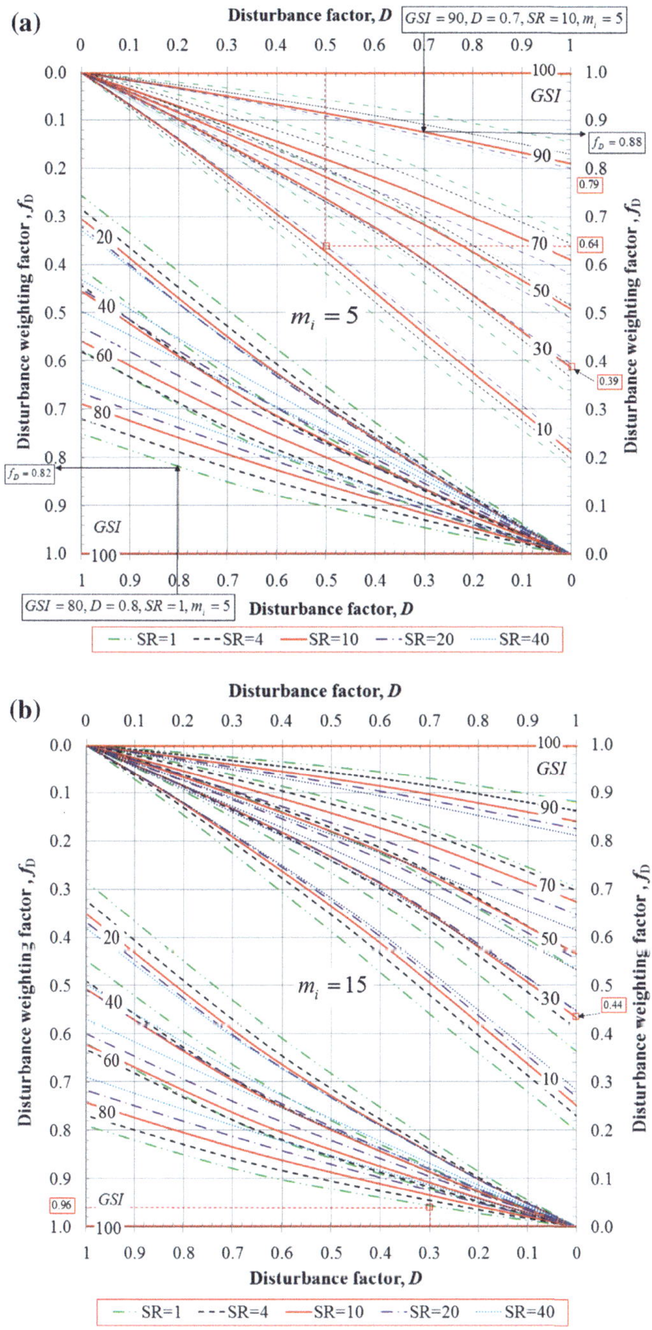

Fig. 11.10 Chart for estimating disturbance weighting factor, f_D (m_i = 5, 15, 25, 35)

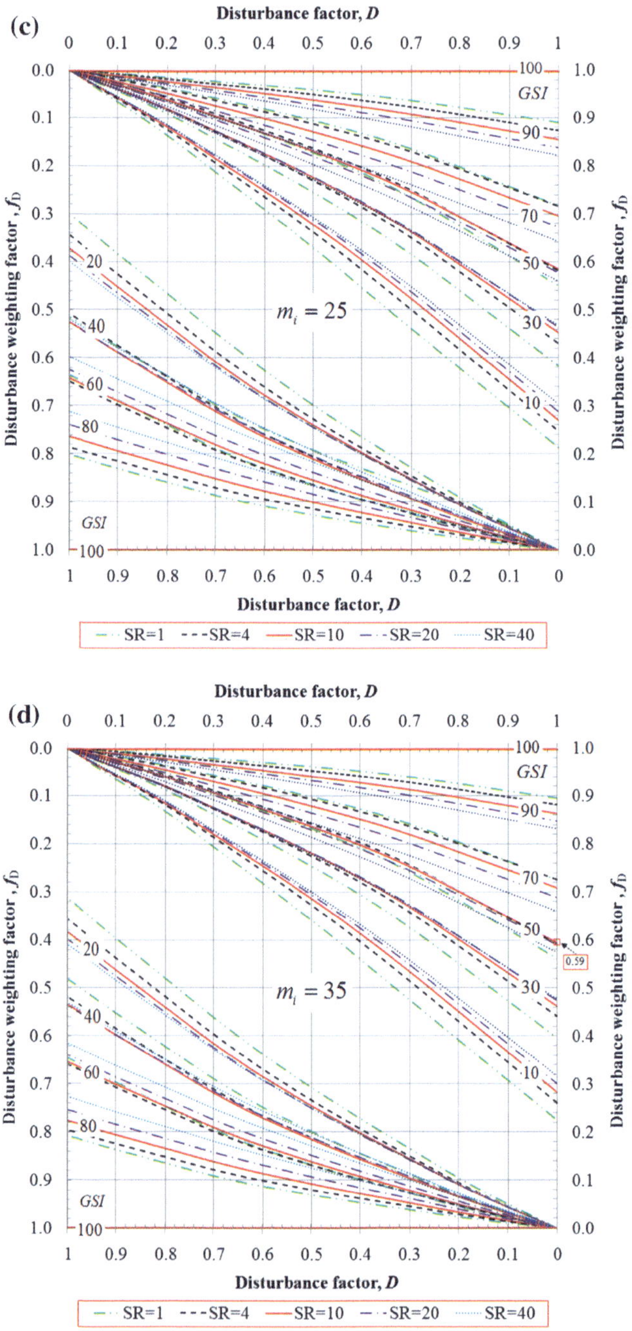

Fig. 11.10 (continued)

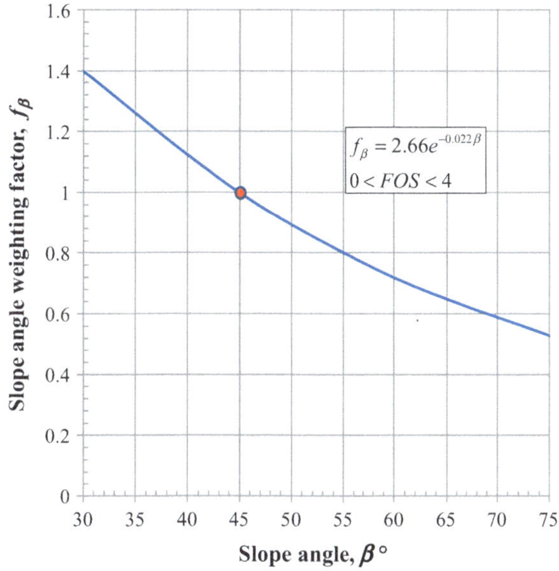

Fig. 11.11 Slope angle weighting factor chart

Compared with the values calculated in *Slide 6.0*, the values of the FOS estimated from Fig. 11.11 show some discrepancy. The prediction performance of Fig. 11.11 was tested using 1680 sets of data. The discrepancy result shows that 78.6% of the data is lower than ±5%, while the absolute average relative error percentage (AAREP) is 3.1%, and the maximum discrepancy percentage (DP_{Max}) is −18.9% as shown in Fig. 11.12. It was also found that the data with a discrepancy greater than ±10% appears when GSI > 90.

Figure 11.11 can also be used in conjunction with the Carranza-Torres (Jimenez et al. 2008) slope stability chart shown in Fig. 11.2, which is based on $\beta = 45°$, for estimating the FOS of a slope with slope angles other than 45°. Example 2 in Table 11.1, therefore, was reanalyzed using the chart from Fig. 11.2 together with f_β from Fig. 11.11. Using the Fig. 11.2, results in the $FOS_{45°} = 0.62$. Using the data from Fig. 11.11, the slope angle weighting factor $f_\beta = 0.72$. Finally, the $FOS = f_\beta \times FOS_{45°} = 0.72 \times 0.62 = 0.446$, which is slightly different from the FOS = 0.489 calculated using *Slide 6.0*.

11.3.5 The Use of the Proposed Stability Charts

The use of the proposed rock slope stability charts to calculate the FOS of a given slope is quite straightforward. Firstly, for given values of SR, GSI and m_i, the value of $FOS_{45°}$ can be obtained using the stability charts (Fig. 11.5). Secondly, for any given disturbance factor D, the disturbance weighting factor f_D can be obtained from Fig. 11.10. Thirdly, for the given slope angle β, the slope angle weighting

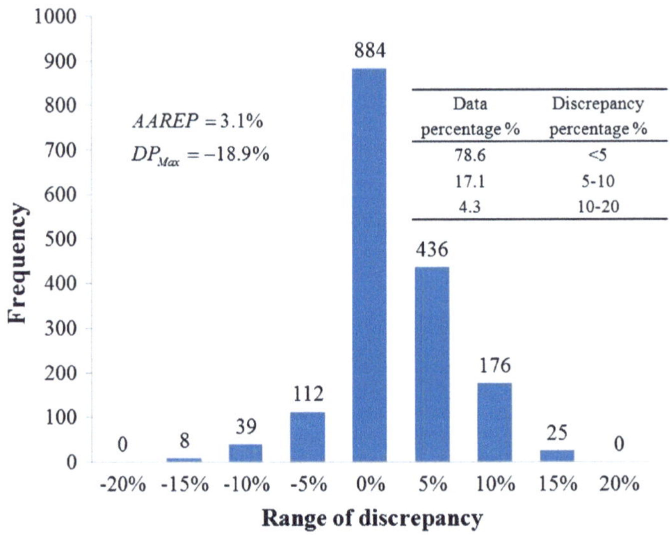

Fig. 11.12 Discrepancy analysis of the proposed rock slope stability charts

factor f_β can be calculated from Eq. (11.14) or obtained from Fig. 11.11. Finally, the FOS can be calculated as, FOS = $f_\beta \times f_D \times \text{FOS}_{45^\circ}$.

Example 2 in Table 11.1 was again adopted to illustrate the use of the proposed charts. The calculation steps are as follows: Firstly, $m_i = 5$ from Fig. 11.5a and $m_i = 10$ from Fig. 11.5b were used to estimate the average value of the FOS for $m_i = 8$. The values of FOS_{45° for $m_i = 5$ and $m_i = 10$ are 1.5 and 1.8, respectively. Therefore, the average value of FOS_{45° for $m_i = 8$ was assumed to equal to 1.65. Then, $m_i = 5$ from Fig. 11.10a and $m_i = 15$ from Fig. 11.10b were used to estimate the average value of f_D for $m_i = 8$. The values of f_D for $m_i = 5$ and $m_i = 15$ are 0.39 and 0.44, respectively. Thus, the value of f_D for $m_i = 8$ should be located between 0.39 and 0.44. Thirdly, slope angle weighting factor f_β for $\beta = 60°$ was estimated using the chart (Fig. 11.11) or Eq. (11.14), with the result $f_\beta = 0.72$. Finally, the lower and upper values of the FOS can be calculated. The results were $\text{FOS}_{\text{Lower}} = f_\beta \times f_{\text{D-Lower}} \times \text{FOS}_{45^\circ} = 0.72 \times 0.39 \times 1.65 = 0.463$ and $\text{FOS}_{\text{Upper}} = f_\beta \times f_{\text{D-Upper}} \times \text{FOS}_{45^\circ} = 0.72 \times 0.44 \times 1.65 = 0.522$. The result provided by *Slide 6.0* was FOS = 0.489.

11.4 Slope Cases Application

The following three examples with a wide range of rock properties and slope geometry were used to illustrate the practical application of the proposed rock slope stability charts. The results are shown in Table 11.5. Example 1: A small slope consisting of highly fractured rock masses with the following input parameters: σ_{ci}

Table 11.5 Three slope examples analyzed using the proposed stability charts

Input parameters	Example 1	Example 2	Example 3
σ_{ci}, MPa	2.7	0.625	46
GSI	10	80	50
m_i	5	15	35
γ, kN/m^3	27	25	23
H, m	5	25	250
β, °	30	75	60
D	0.5	0.3	1
Calculated data			
SR: $\sigma_{ci}/\gamma H$	20	1	8
$FOS_{45°}$	1.1	2.08	3.3
f_D	0.64	0.96	0.59
f_β	1.4	0.53	0.72
Factor of safety			
Proposed charts	0.986	1.058	1.402
Slide 6.0	1.025	1.045	1.391
Discrepancy (%)	−3.84	1.27	0.78

= 2.7 MPa, GSI = 10, $m_i = 5$, $\gamma = 27\text{kN/m}^3$, $H = 5$ m and $\beta = 30°$, $D = 0.5$. Example 2: A medium slope consisting of good quality rock masses with the following input parameters: $\sigma_{ci} = 0.625$ MPa, GSI = 80, $m_i = 15$, $\gamma = 25\text{kN/m}^3$, $H = 25$ m and $\beta = 75°$, $D = 0.3$. Example 3: A large open pit slope consisting of blocky rock masses with the following input parameters: $\sigma_{ci} = 46$ MPa, GSI = 50, $m_i = 35$, $\gamma = 23\text{kN/m}^3$, $H = 250$ m and $\beta = 60°$, $D = 1.0$. The results show that there is close agreement between the proposed stability chart and the Slide 6.0 results. The discrepancy percentages for Examples 1–3 are −3.84%, 1.27% and 0.78%, respectively.

11.5 Conclusions

New rock slope stability charts for estimating of the stability of rock mass slopes satisfying with the GHB criterion have been proposed. The proposed charts can be used to calculate the FOS of a slope directly from the Hoek–Brown parameters (GSI, mi and D), slope geometry (β and H) and rock mass properties (σ_{ci} and γ).

Firstly, the theoretical relationship between the strength ratio (SR), $\sigma_{ci}/(\gamma H)$ and the FOS has been demonstrated. It is found that when the values of β, GSI, m_i and D in a homogeneous slope are given, the FOS of a slip surface for a particular method of slices is uniquely related to the parameter SR regardless of the magnitude of the individual parameters σ_{ci}, γ and H. Based on the relationship between the SR and FOS, stability charts as shown in Fig. 11.5 for calculating the FOS of a slope with specified slope angle $\beta = 45°$, $D = 0$ have been proposed.

Secondly, while the disturbance factor D has great influence upon the stability of rock mass slopes, it is, nevertheless, difficult to determine its exact value. Yet a sensitivity analysis of D is probably more significant in judging the acceptability of a slope design than a single calculated FOS with specified D values estimated from the guidelines by Hoek and Brown (1980), Hoek and Diederichs (2006), Sonmez et al. (2006). We proposed a disturbance weighting factor f_D as shown in Fig. 11.10 to show the influence of a range of values of D upon the stability of rock mass slopes.

Thirdly, a slope angle weighting factor f_β has been proposed to show the influence of the slope angle β on slope stability. It should be noted that the chart, as shown in Fig. 11.11, representing the relationship between f_β and β was proposed based on the data $0 < \mathrm{FOS} < 4$, however, it will be applicable for most civil and mining slope cases.

Combined with stability charts based on $\beta = 45°$, the values of f_β and f_D can be used for estimating the FOS of slopes with various angles in a variety of blast damage and stress relief conditions. The reliability of the proposed charts has been tested against results from *Slide 6.0* using 1680 sets of data representing a wide range of rock mass properties and slope geometries. The results show that there is good agreement between the values of the FOS as calculated from the charts and those produced by *Slide 6.0*, as shown in Fig. 11.12. The discrepancy of 78.6% of data is lower than ±5%, and the absolute average relative error percentage (AAREP) is 3.1%. In addition, it is found that the data with a discrepancy of more than ±10% appear when GSI > 90.

The proposed charts are quite simple and straightforward to use and can be adopted as a useful tool for the preliminary rock slope stability analysis.

References

Baker R (2003) A second look at Taylor's stability chart. J Geotech Geoenviron Eng 129 (12):1102–1108

Baker R, Shukha R, Operstein V, Frydman S (2006) Stability charts for pseudo-static slope stability analysis. Soil Dynam Earthq Eng 26:813–823

Cai M, Kaisera PK, Uno H, Tasaka Y, Minami M (2004) Estimation of rock mass deformation modulus and strength of jointed hard rock masses using the GSI system. Int J Rock Mech Min Sci 41:3–19

Carranza-Torres C (2004) Some comments on the application of the Hoek–Brown failure criterion for intact rock and rock masses to the solution of tunnel and slope problems. In: Barla G, Barla, M (eds) MIR 2004—X conference on rock and engineering mechanics, Torino, 24–25 November 2004, pp 285–326

Hoek E (2007) Rock mass properties. Practical Rock Engineering. http://www.rocscience.com/hoek/pdf/11_Rock_mass_properties.pdf

Hoek E, Brown E (1997) Practical estimates of rock mass strength. Int J Rock Mech Min Sci 34 (8):1165–1186

Hoek E, Brown ET (1980) Underground excavations in rock. Instn Min Metall, London

Hoek E, Bray JW (1981) Rock slope engineering, 3rd edn. Instn Min Metall, London

Hoek E, Carranza-Torres C, Corkum B (2002) Hoek–Brown failure criterion—2002 Edition. In: Proceedings of NARMS-TAC. Mining Innovation and Technology, Toronto

Hoek E, Diederichs MS (2006) Empirical estimation of rock mass modulus. Int J Rock Mech Min Sci 43:203–215

Jimenez R, Serrano A, Olalla C (2008) Linearization of the Hoek and Brown rock failure criterion for tunnelling in elasto-plastic rock masses. Int J Rock Mech Min Sci 45:1153–1163

Li AJ, Cassidy MJ, Wang Y, Merifield RS, Lyamin AV (2012) Parametric Monte Carlo studies of rock slopes based on the Hoek–Brown failure criterion. Comput Geotech 45:11–18

Li AJ, Lyamin AV, Merifield RS (2009) Seismic rock slope stability charts based on limit analysis methods. Comput Geotech 36:135–148

Li AJ, Merifield RS, Lyamin AV (2008) Stability charts for rock slopes based on the Hoek–Brown failure criterion. Int J Rock Mech Min Sci 45(5):689–700

Li AJ, Merifield RS, Lyamin AV (2011) Effect of rock mass disturbance on the stability of rock slopes using the Hoek–Brown failure criterion. Comput Geotech 38:546–558

Muirwood AM (1972) Tunnels for road and motorways. Q J Eng Geol 5:111–126

Michalowski RL (2010) Limit analysis and stability charts for 3D slope failures. J Geotech Geoenviron Eng 136:583–593

Naghadehi M, Jimenez R, KhaloKakaie R, Jalali S (2013) A new open-pit mine slope instability index defined using the improved rock engineering systems approach. Int J Rock Mech Min Sci 61:1–14

Priest SD (2005) Determination of shear strength and three-dimensional yield strength for the Hoek–Brown criterion. Rock Mech Rock Eng 38(4):299–327

Phase^2 8.0. www.rocscience.com

RocData 4.0. www.rocscience.com

Shen J, Priest SD, Karakus M (2012a) Determination of Mohr-Coulomb shear strength parameters from Generalized Hoek–Brown criterion for slope stability analysis. Rock Mech Rock Eng 45:123–129

Shen J, Karakus M, Xu C (2012b) Direct expressions for linearization of shear strength envelopes given by the Generalized Hoek–Brown criterion using genetic programming. Comput Geotech 44:139–146

Sheorey PR (1997) Empirical rock failure criteria. Balkema, Rotterdam

Slide 6.0. www.rocscience.com

Steward T, Sivakugan N, Shukla SK, Das BM (2011) Taylor's slope stability charts revisited. Int J Geomech 11(4):348–352

Sonmez H, Gokceoglu C (2006) Discussion of the paper by E. Hoek and M.S. Diederichs "Empirical estimation of rock mass modulus". Int J Rock Mech Min Sci 43:671–76

Taylor DW (1937) Stability of earth slopes. J Boston Soc Civil Eng XXIV(3):337–386

Taheri A, Tani K (2010) Assessment of the stability of rock slopes by the slope stability rating classification system. Rock Mech Rock Eng 43:321–333

Wyllie DC, Mah C (2004) Rock slope engineering: civil and mining, 4th edn. Spon Press, New York

Zanbak C (1983) Design charts for rock slopes susceptible to toppling. J Geotech Eng Div ASCE 190(8):1039–1062

Chapter 12
The Effects of Blast Damage Zone Thickness on Rock Slope Stability

Abstract The selection of blast damage zone T of the Hoek–Brown (HB) criterion is significant in open pit slope design and stability analysis. Traditional slope numerical analysis adopts a single value of blast damage factor D to the entire rock mass, which results in underestimation of slope stability. In this research, the parallel layer model (PLM), which rock mass is divided into a number of layers parallel to slope surface with a decreasing value of D applied to each layer in slope modeling, is used in the limit equilibrium method to investigate the effect of blast damage zone T on the stability of rock slopes. Based on extensive parametric studies, a blast damage zone weighting factor f_T is proposed to quantify the influence of T on the evaluation of the factor of safety (FOS) of given slopes. Results show that the selection of T in the slope model plays an important role in terms of calculation of FOS, especially, when the ratio of T to slope height H is less than 1.0. Finally, based on f_T and the existing stability charts, a stability model is proposed for estimating the FOS of slopes with various slope geometries and rock mass properties. The reliability of the proposed stability model is tested against numerical solutions. The results show that FOS estimated from the proposed stability model exhibits only 5.6% average discrepancy from numerical solutions using 1254 sets of data. The proposed stability model is simple and straightforward, and can be used for the preliminary rock slope stability analysis considering blast damage zone effects.

12.1 Introduction

Practical experience in the design of rock slope projects has demonstrated that the estimated rock mass strength of a given slope is affected by the blast damage in slope excavations (Hoek and Brown 1988). In order to improve the accuracy of rock mass strength prediction under blast conditions, a blast damage factor D, also known as the disturbance factor, was introduced in the Hoek–Brown criterion (see Eq. 12.1) to calculate the HB input parameters (Hoek et al. 2002).

J. Zuo and J. Shen, *The Hoek-Brown Failure criterion—From theory to application*,
https://doi.org/10.1007/978-981-15-1769-3_12

$$\sigma_1 = \sigma_3 + \sigma_{ci}(m_b \frac{\sigma_3}{\sigma_{ci}} + s)^a \tag{12.1}$$

where σ_1 and σ_3 are the maximum and minimum principal stresses at failure, respectively; σ_{ci} is the uniaxial compressive strength (UCS) of the intact rock. m_b, s and a are the HB input parameters which can be estimated from the HB constant m_i of intact rocks, the Geological Strength Index (GSI) and the blast damage factor D, given by:

$$m_b = m_i \exp\left(\frac{GSI - 100}{28 - 14D}\right) \tag{12.2}$$

$$s = \exp\left(\frac{GSI - 100}{9 - 3D}\right) \tag{12.3}$$

$$a = \frac{1}{2} + \frac{1}{6}(e^{-GSI/15} - e^{-20/3}) \tag{12.4}$$

The value of m_i depends on rock types, mineral compositions, grain sizes and cementation of rocks, and it can be estimated based on regression analyses of triaxial experimental data of intact rocks. GSI depends on rock mass structures and surface conditions. The details of calculation and selection of m_i and GSI can be found in the papers by Hoek and Brown (1997), Sari (2010), Cai et al. (2004), Marinos et al. (2005), Shen and Karakus (2014a), Read and Richards (2014), Zuo et al. (2015), Bertuzzi et al. (2016), Vásárhelyi et al. (2016), Wang and Shen (2017), Morelli (2017), Marinos and Carter (2018) and Hoek and Brown (2018).

The value of D varies from zero for undisturbed rock masses to one for highly disturbed rock masses according to disturbance conditions caused by blast damage as well as stress relief. The selection of a suitable value of D and a blast damage zone thickness T to quantify appropriate blast damage effects in the numerical modeling is a very important issue for the slope stability analysis.

Engineering experience in the design of large open pit slopes has demonstrated that it is not easy to estimate the accurate value of D as various factors can influence the degree of disturbance in the rock mass. Hoek et al. (2002), Hoek and Diederichs (2006) and Hoek (2007), therefore, presented a number of slope cases to illustrate how to select reasonable values of D for practical projects. In civil engineering applications, small scale slope blasting results in modest rock mass damage, therefore $D = 1.0$ and 0.7 are suggested under poor and good blasting conditions, respectively. For folded sedimentary rock slope in a carefully excavated road cutting, $D = 0.3$ is suggested. In mining applications, large scale open pit slopes suffer significant disturbance under heavy production blasting, $D = 1.0$ is the suggested value. For softer rocks under mechanical excavation, $D = 0.7$ is suggested.

The thickness of blast damage zone depends on blast design factors (explosive type, charge, drill hole diameter and burden) and rock mass mechanical properties.

Lupogo (2017) noted that the persistence of pre-existing discontinuities could significantly influence the thickness of blast damage zone. The rock mass with a lower discontinuity persistence tends to develop a low thickness of blast damage zone compared to a rock mass with a high discontinuity persistence. Furthermore, it was observed that the blast damage zone thickness decreases with an increase in the tensile strength of the rock mass. Mohanty and Chung (1986) divided the blast damage zone into two types in open pit mines: near field damage which is caused by high frequency vibration when the explosion occurs in the vicinity of the slope wall (0–20 m), and far field (up to 100 m) damage which is caused by low frequency vibration. It should be noted that backbreak, which can be defined as overbreak of rocks beyond the limits of the rear row of holes in a blast pattern, could be observed in the blast damage zone due to inappropriate blast design. This phenomenon can cause problems of mine wall instability and improper fragmentation, and, consequently, increase the cost of a mining operation.

Hoek (2012) presented some preliminary guidelines as a starting point for the determination of the thickness of the blast damage zone as a result of production blast. For example, for a large production blast, $T = 2.0$ to $2.5\ H$ when the blast is confined with little or no control; $T = 1.0$ to $1.5\ H$ when the blast is to a free face with no control; and $T = 0.5$ to $1.0\ H$ when the blast is to a free face with some controls, e.g. one or more buffer rows used.

The determination of the rate of reduction of D in blast damage zone thickness T is another important task, which should be considered in the numerical modelling for the slope stability analysis. Currently, it is still not clear about the pattern of damage reduction with depth behind the slope face. Li et al. (2011) and Qian et al. (2017) adopted a linear decrease of D from the slope face to the back of the slope. Lupogo (2017) carried out numerical simulation and found that the degree of damage has a negative exponential correlation with depth behind the slope face.

These guidelines for the selection of D and T are based on a limited number of case histories, and it can be argued that they should be extended and modified by considering more practical cases. Therefore, in order to understand the actual effects of D and T upon the rock slope stability, it is critical that engineers and researchers perform studies based on a range of D and T values rather than relying on the results from a single set of values.

Currently, there are three models available for the slope stability analysis considering the effects of blasting damage on the calculation of factor of safety (FOS) of given slopes, as shown in Fig. 12.1.

Based on the literature review, it is found that the traditional slope model (Fig. 12.1a) is still used in many current researches (Shen et al. 2013; Shen and Karakus 2014b; Jiang et al. 2016; Deng et al. 2016; Sun et al. 2016), where the blast damage zone is not considered and a single value of D for the entire rock mass slopes is adopted. Figure 12.1b shows the modified slope model which adopts a single value of D only for the blast damage zone in the slope, as used in Turkey (2012). As suggested by Hoek and Karzulovic (2000), however, the stability of disturbed slopes is controlled by the properties of the blast damaged rock mass. Therefore, to have a more realistic model, it is recommended that a rock mass slope

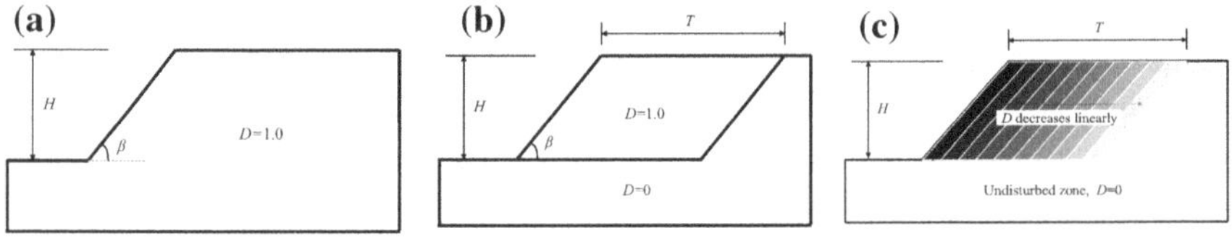

Fig. 12.1 Diagrammatic representation of different slope models considering blasting effects **a** the traditional slope model that adopts a single value of D to the entire rock slope, **b** the modified slope model that adopts a single value of D only for the blast damage zone, **c** the parallel layer model (PLM) that adopts a linear decreasing D assigned to each layer in the blast damage zone

should be divided into a number of layers with a decreasing value of D applied to each layer in slope models. Here, such a model is termed the parallel layer model (PLM) in this work, as shown in Fig. 12.1c. The damage in the disturbed rock zone declines from severe to slight with increasing depth into the undisturbed rock mass. Similar PLM models were used successfully in Li et al. (2011), Qian et al. (2017) and Yilmaz et al. (2018) for their rock slope stability analysis.

Rock slope stability assessment is critical in open pit mines, and the HB criterion has been successfully used for the estimation of rock mass strength in rock engineering for the past decades (Sari 2012; Deng et al. 2016; Melentijevic et al. 2017; Peng et al. 2017). Most slope stability analyses are formulated to calculate the FOS which is a common measure of the safety margin of a given slope. Due to the convenience of quick assessment of preliminary slope designs, stability charts have been extensively used for assessing the slope stability in practical engineering (Li et al. 2008; Michalowski 2010; Shen et al. 2013; Wan et al. 2016; Qian et al. 2017).

As input parameters of the HB criterion can be obtained directly from assessment of uniaxial and triaxial compressive testing of intact rocks, measurement of discontinuity characteristics of rock masses and evaluation of blast conditions, many stability charts were developed to estimate the FOS directly from the HB input parameters for the past ten years using different analysis methods, such as limit analysis method (Li et al. 2008, 2009), limit equilibrium method (Shen et al. 2013; Jiang et al. 2016) and finite element method (Sun et al. 2016).

However, these stability charts are based on the traditional blast damage model (see Fig. 12.1a) which does not consider the effects of blast damage zone thickness T on the slope stability. Li et al. (2011) used a model similar to the parallel layer model to assess the stability of slope using limit analysis method where the blast damage zone thickness T is set to slope height H. They concluded that using constant or varying D distributions would lead to very different values of FOS. The PLM was further used in Qian et al. (2017) by taking into account various values of T, ranging from 0.5 to 2.5 H. They found that the value of T has great influence on the rock slope stability.

In this Chapter the parallel layer model (PLM) is used with the limit equilibrium method (LEM) to investigate the effects of blast damage zone thickness T on the stability of open pit slopes. Based on extensive parametric studies, a blast damage zone thickness weighting factor f_T is proposed in this study to quantify the influence of T on the FOS of given slopes. The weighting factor f_T is then combined with existing stability charts based on the HB criterion to form a proposed stability model for the estimation the FOS of slopes with different slope geometries and rock mass properties.

12.2 Parallel Layer Model (PLM)

In this research, the parallel layer model, which assigns a linearly decreasing D to different layers in the slope, was adopted for stability analysis using *Slide 6.0* program. The blast damage zone in the slope model is divided into a number of layers with changing rock material properties, as shown in Fig. 12.1c. The value of D in each layer decreases linearly from slope face at $D = 1$ to the deep undisturbed rock mass at $D = 0$. The values of other input rock mass parameters, γ, GSI, m_i and σ_{ci} in each layer are identical.

It should be noted that the number of layers of blast damage zone could influence the FOS in LEM, as shown in Table 12.1.

Table 12.1 shows that the value of FOS tends to be stabilizing when n is above 14. As the maximum number of material types is 20 in *Slide* 6.0, therefore, in subsequent analyses, n is set to 19 for the disturbed rock masse, and the last one is used for the undisturbed rock mass.

Table 12.1 Comparison of FOS of rock slopes with different number of layers in PLM

σ_{ci} (MPa)	γ (kN/m^3)	H (m)	β (°)	GSI	m_i	D	T/H	Number of layers, n	FOS
2.7	27	10	45	50	20	1.0	2.5	1	1.79
2.7	27	10	45	50	20	1.0	2.5	4	1.79
2.7	27	10	45	50	20	1.0	2.5	6	1.79
2.7	27	10	45	50	20	1.0	2.5	7	1.79
2.7	27	10	45	50	20	1.0	2.5	8	1.80
2.7	27	10	45	50	20	1.0	2.5	10	1.83
2.7	27	10	45	50	20	1.0	2.5	12	1.88
2.7	27	10	45	50	20	1.0	2.5	13	1.90
2.7	27	10	45	50	20	1.0	2.5	14	1.91
2.7	27	10	45	50	20	1.0	2.5	16	1.91
2.7	27	10	45	50	20	1.0	2.5	18	1.91
2.7	27	10	45	50	20	1.0	2.5	19	1.91

12.3 Comprehensive Stability Analysis Based on PLM

It is not a simple task to determine the exact blast damage zone thickness T in a slope, as various factors, such as rock properties and blast design, can influence the degree of disturbance in the rock mass. As noted by Hoek and Diederichs (2006), the sensitivity analysis of a design is probably more useful in judging the acceptability of the design than a single calculated FOS. Thus, in order to understand the actual influence of T upon the stability of slopes, it is critical that researchers and engineers perform studies based on a range of T values rather than relying on the results from a single T value.

Here, based on the parallel layer model, limit equilibrium analyses were carried out to investigate the effects of T on the stability of rock slopes. The work outlined here requires hundreds of runs on a computer for evaluating FOS of slopes having different slope geometries and rock mass properties. Details of the slope model are the same as those used in Shen et al. (2013). $D = 1.0$ is assigned to the slope face, which represents large scale open pit mine slopes with a significant disturbance from heavy production blasts. The blast damage zone thickness T is assigned from 0.5 H to 2.5 H.

The strength ratio (SR) in Fig. 12.2, expressed as the ratio of σ_{ci} (uniaxial compressive strength) to γH (vertical stress of the rock slope), is a dimensionless parameter. Shen et al. (2013) demonstrated that when the values of β, GSI, m_i and D in a homogeneous slope are given, the value of FOS of a failure surface for a particular method of slices is only related to the parameter SR regardless of the values of individual parameters σ_{ci}, unit weight of rock masses γ and slope height H. Figure 12.2 presents the parametric results based on the HB criterion with a range of SR, GSI, and m_i, but with a specified slope angle $\beta = 45°$. It should be noted that $T = \infty$ represents the traditional model which assigns a single value of D to the entire slope. The dash lines in Fig. 12.2 represent GSI values less than 40. In general, blasting is unnecessary in excavating rock masses with GSI < 40 (Tsiambaos and Saroglou 2010).

12.4 The Blast Damage Zone Thickness Weighting Factor f_T

Here, we proposed a blast damage zone thickness weighting factor f_T which is the ratio of FOS_{PLM} (FOS calculated from the PLM) to FOS_{∞} (FOS calculated from the traditional model). $f_T = FOS_{PLM}/FOS_{\infty}$ reflects the effects of T on the stability of rock slopes. As an example, for a given slope with $\sigma_{ci} = 1080$ kPa, $\gamma = 27$ kN/m^3, $H = 10$ m, GSI = 60 and $m_i = 15$, the FOS corresponding to different T/H conditions were calculated, and the results are shown in Table 12.2. For this example, FOS is 1.58 for $T/H = \infty$ and FOS is 2.05 for $T/H = 0.5$ and therefore, $f_T = FOS_{PLM}/FOS_{\infty} = 2.05/1.58 = 1.30$. Comprehensive results for the estimation of f_T are shown in Fig. 12.3. It is found that there is an obvious trend of decrease of f_T with the increase of GSI and T/H.

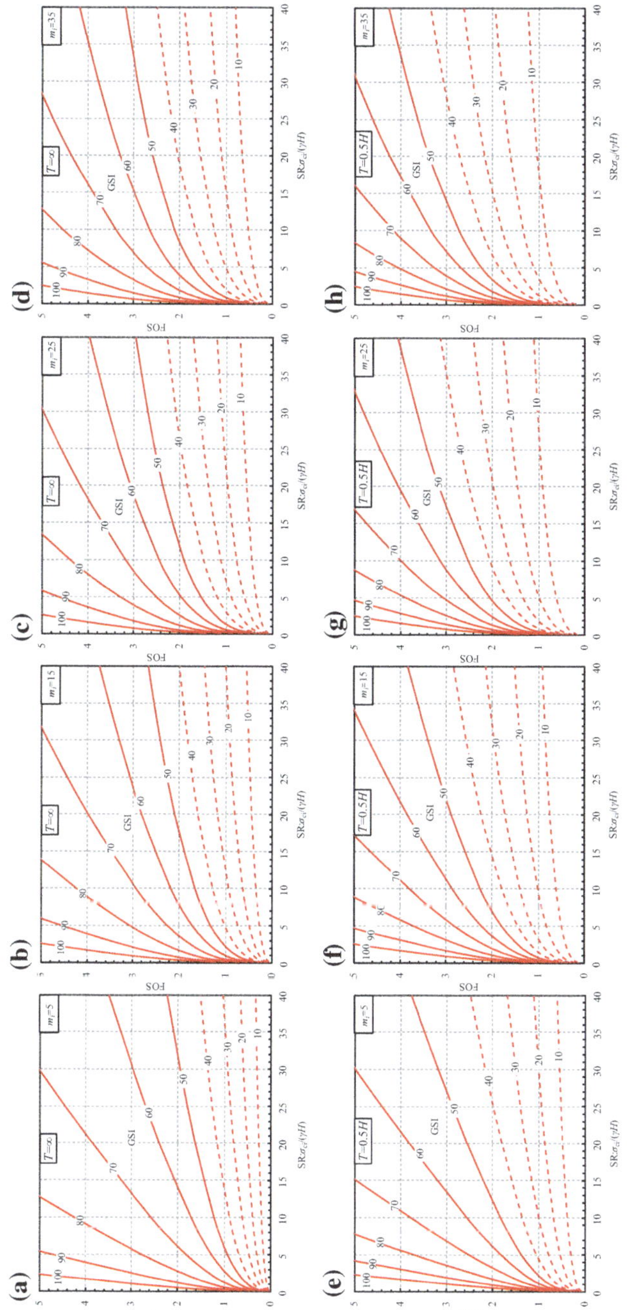

Fig. 12.2 Stability charts for rock mass slope based on PLM, $\beta = 45°$

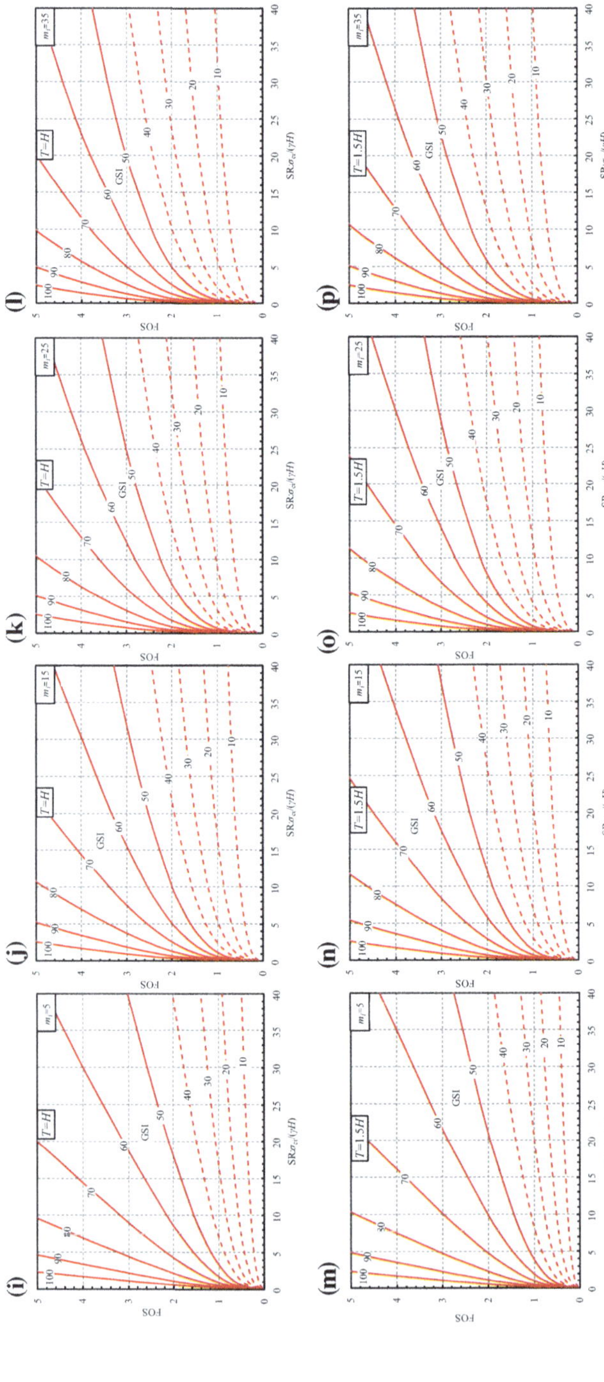

Fig. 12.2 (continued)

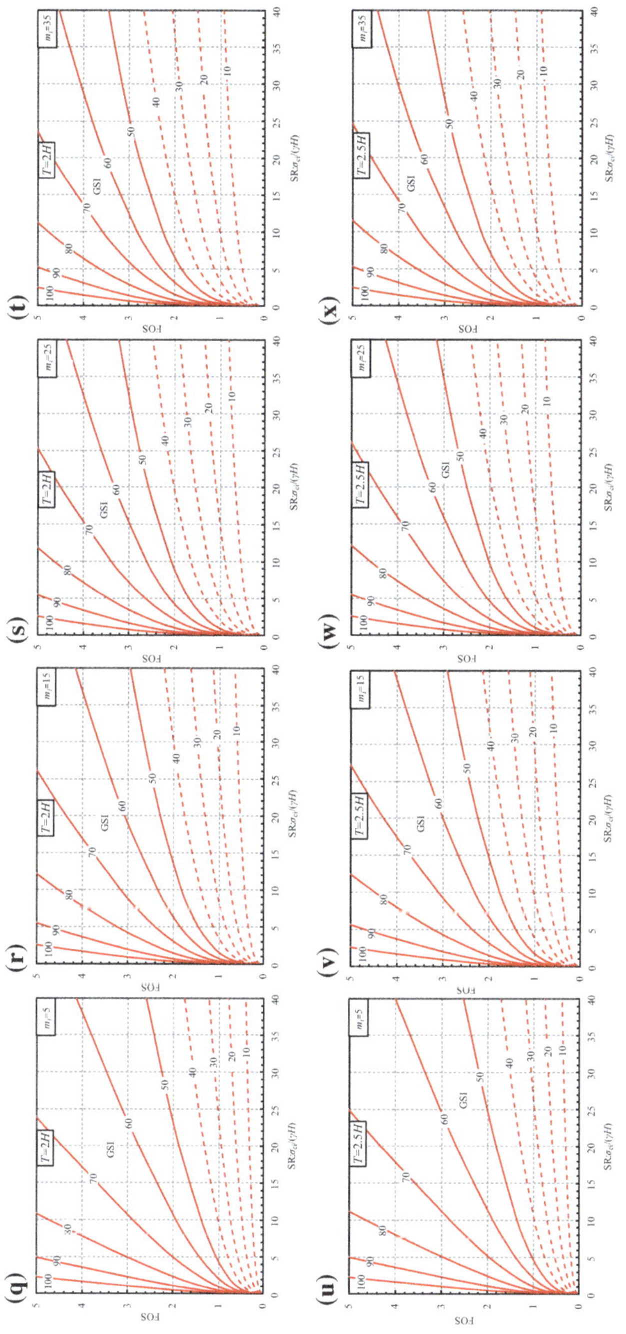

Fig. 12.2 (continued)

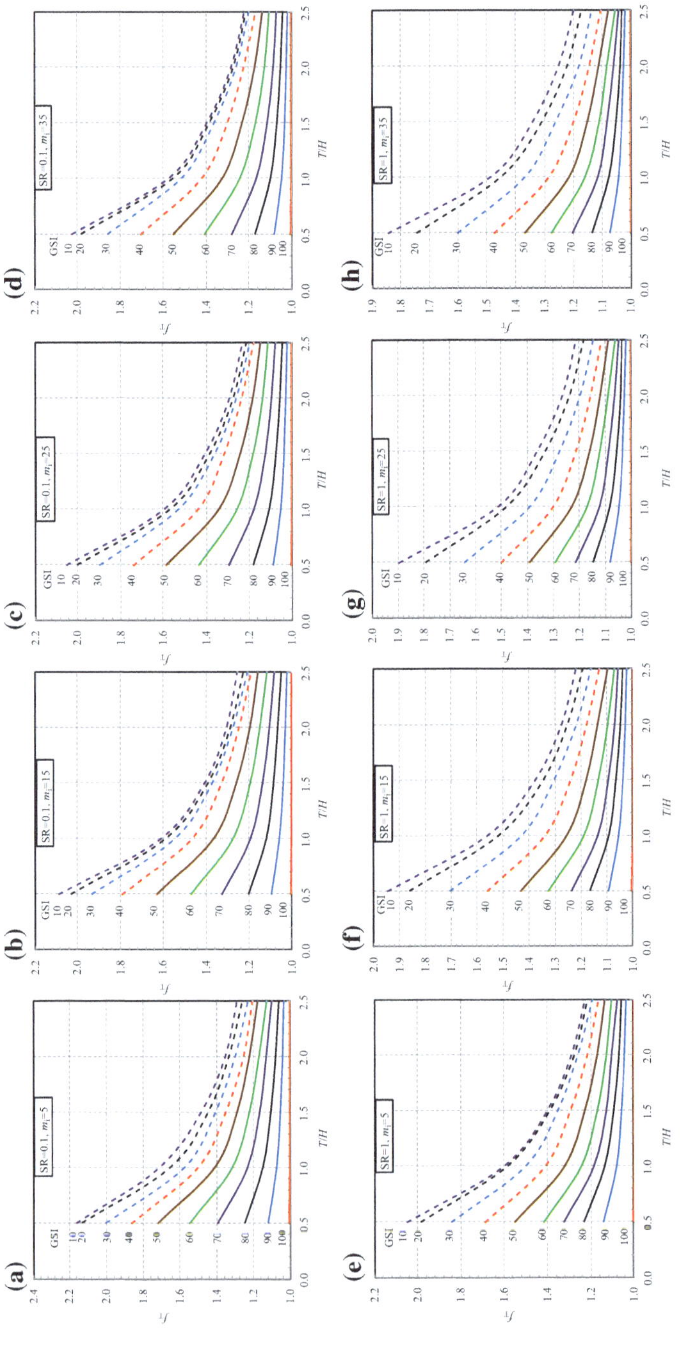

Fig. 12.3 Charts for estimating blast damage zone thickness weighting factor f_T

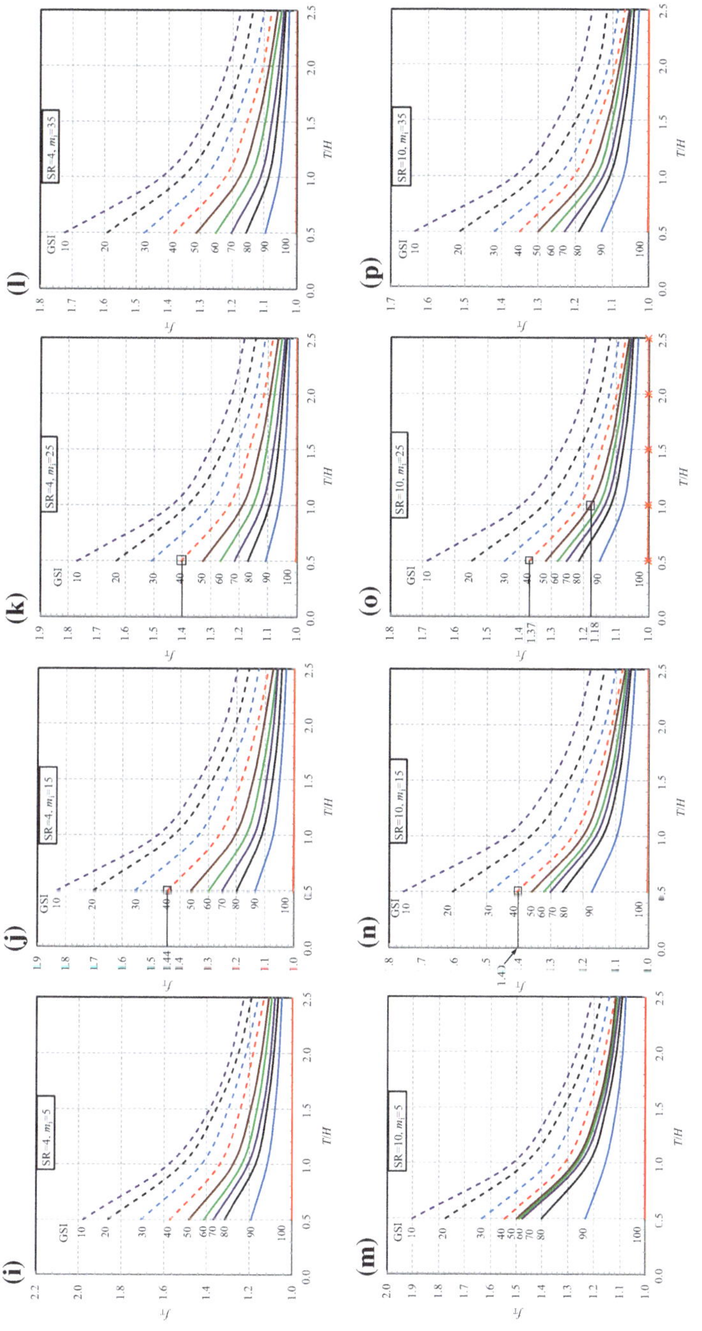

Fig. 12.3 (continued)

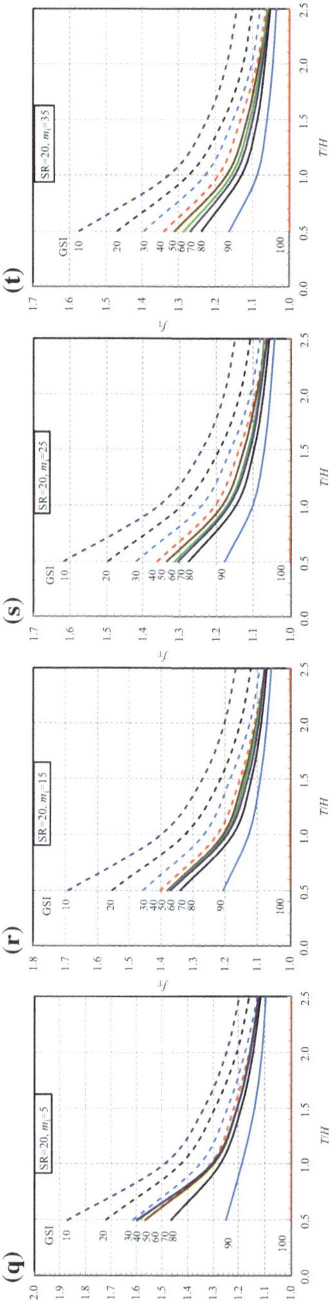

Fig. 12.3 (continued)

Table 12.2 An example for the calculation of f_T

T/H	FOS	f_T
∞	1.58	1.58/1.58 = 1.00
0.5	2.05	2.05/1.58 = 1.30
1.0	1.84	1.84/1.58 = 1.16
1.5	1.76	1.76/1.58 = 1.11
2.0	1.71	1.71/1.58 = 1.08
2.5	1.67	1.67/1.58 = 1.06

Figures 12.4, 12.5 are alternative forms of charts to illustrate the effects of m_i and SR values on f_T, respectively. Figure 12.4 shows that the value of f_T is only slightly influenced by the m_i values, especially when $T/H > 1.0$. It is also found that SR has a considerable effect on f_T when SR < 2.

Based on the statistical analysis of Figs. 12.3, 12.4 and 12.5, it is clear that the value of f_T mainly depends on GSI and T/H. Therefore, a guideline was summarized in Table 12.3 as a starting point for the estimation of f_T for a given slope with different T values. It shows that the selection of T in the slope model plays an important role in the calculation of FOS, especially when $T/H < 1.0$. However, the effect of T on the stability of slope is relatively insignificant when $1.0 < T/H < 2.5$. Similar conclusions were also obtained by Qian et al. (2017). They stated that the blast damage zone thickness only has a small effect on the assessment of rock slope stability when $1.0 < T/H < 2.5$.

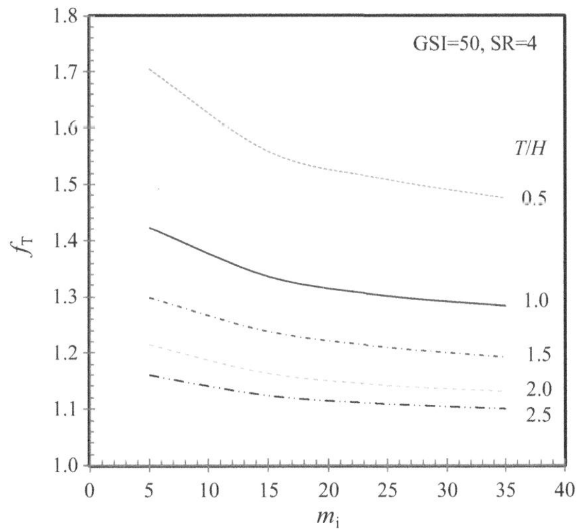

Fig. 12.4 Relationship between f_T and m_i

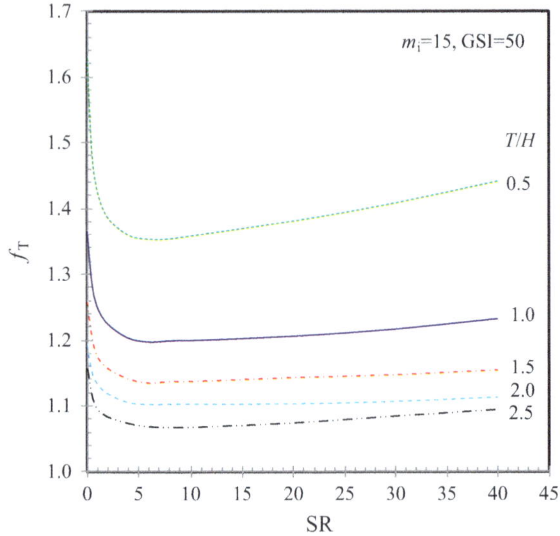

Fig. 12.5 Relationship between f_T and SR

Table 12.3 Guidelines for the estimation of the weighting factor f_T

T/*H*	GSI					
	50	60	70	80	90	100
0.5	1.3–1.7	1.2–1.7	1.2–1.7	1.1–1.5	1.1–1.3	1.0
1.0	1.1–1.4	1.1–1.4	1.1–1.4	1.1–1.3	1.0–1.2	1.0
1.5	1.1–1.3	1.1–1.3	1.1–1.3	1.0–1.2	1.0–1.2	1.0
2.0–2.5	1.1–1.2	1.1–1.2	1.0–1.2	1.0–1.2	1.0–1.1	1.0

12.5 Stability Model Based on f_T and Existing Stability Charts

Based on the weighting factor f_T together with the stability charts proposed by Shen et al. (2013), a stability model (see Eq. 12.5) was proposed for estimating the FOS of slopes with different slope geometries and rock mass properties.

$$\mathrm{FOS_{PLM}} = \mathrm{FOS}_{45^\circ\infty} \times f_D \times f_T \times f_\beta \tag{12.5}$$

where, $\mathrm{FOS}_{45^\circ\infty}$ represents the FOS calculated from traditional model with the specified slope angle of β = 45° and D = 0, as shown in Fig. 12.6. f_D is a disturbance weighting factor which represents the effects of D on the stability of slopes, and can be obtained from Fig. 12.7. f_T is the blast damage zone thickness weighting factor discussed above which represents the influence of T on the stability of slopes and can be estimated from Fig. 12.3. f_β is a slope angle weighting factor which represents the effects of slope angle β on the slope stability. We re-examined the

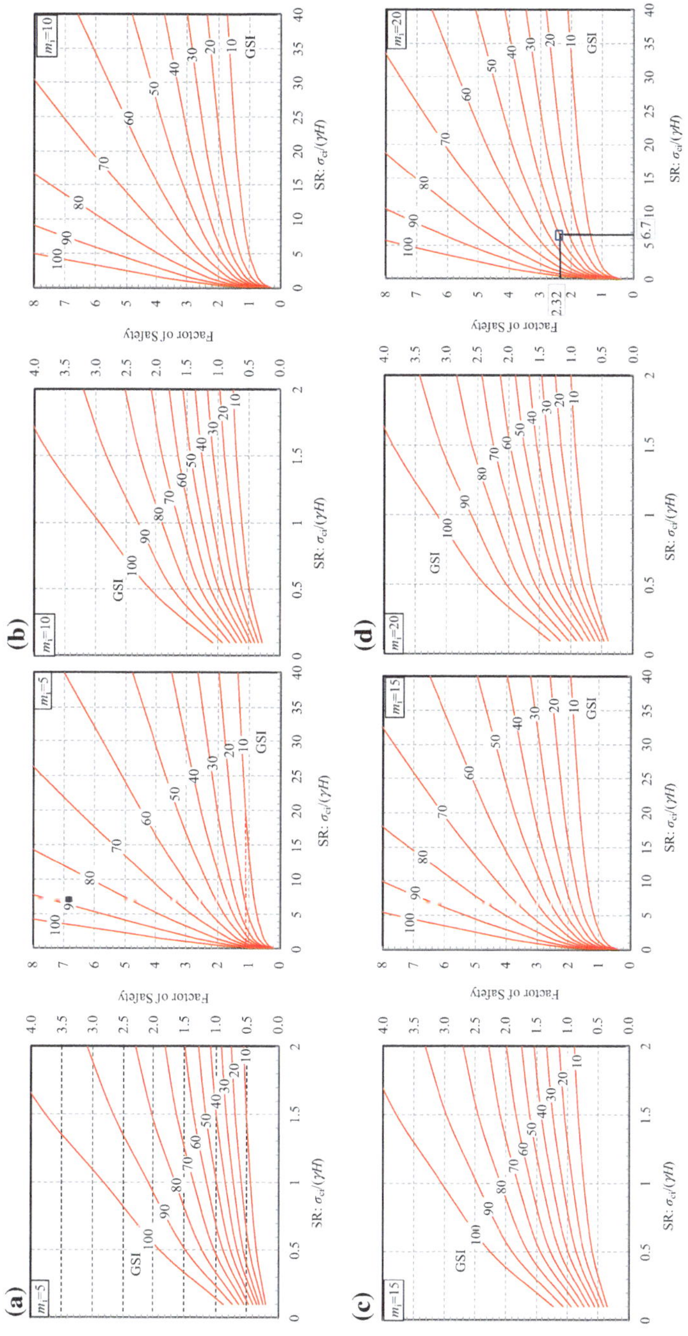

Fig. 12.6 Slope stability charts with $\beta = 45°$, $D = 0$ (Shen et al. 2013)

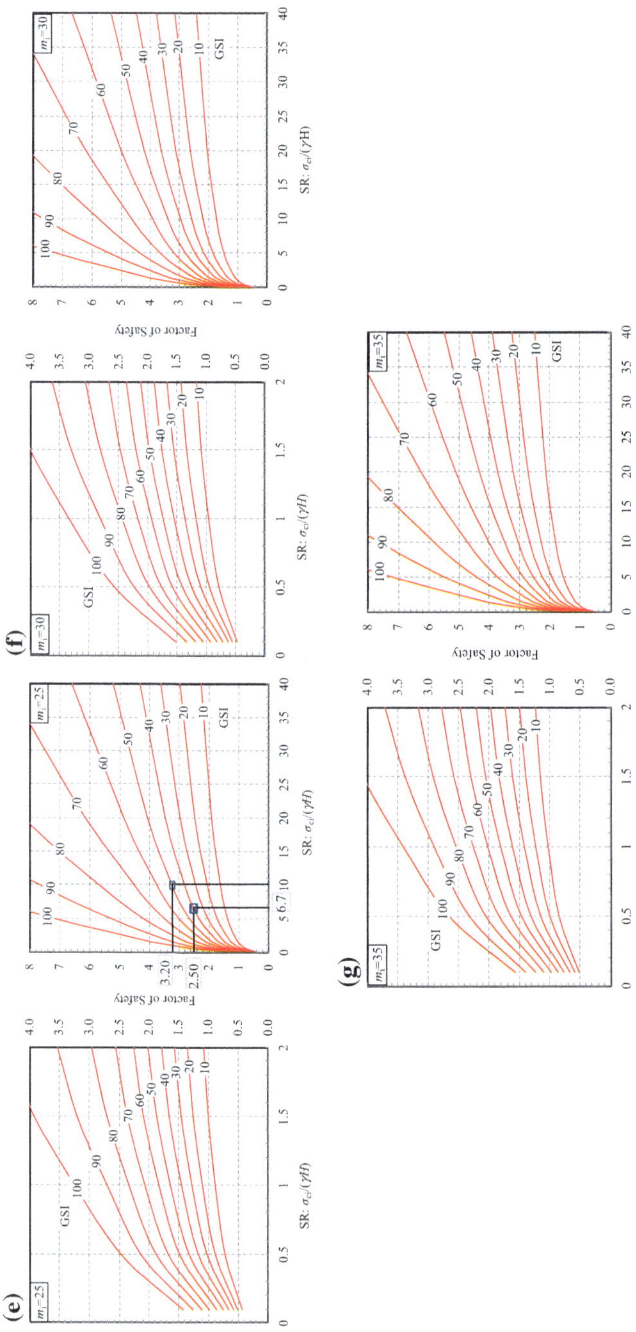

Fig. 12.6 (continued)

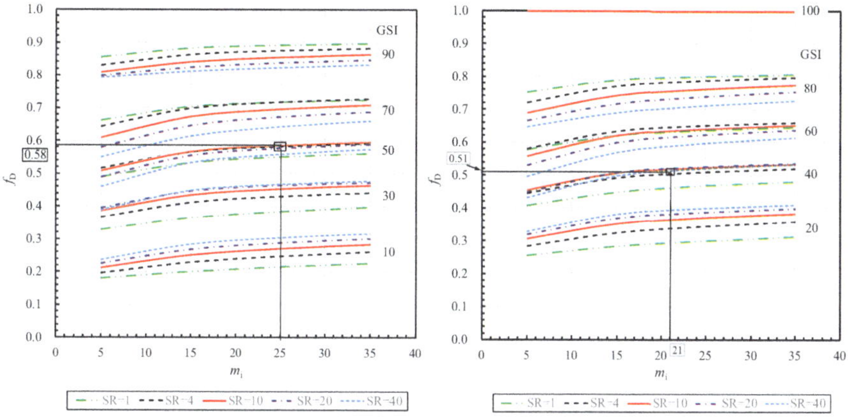

Fig. 12.7 Charts for estimating disturbance weighting factor f_D (after Shen et al. 2013)

Fig. 12.8 Chart for estimating slope angles weighting factor f_β

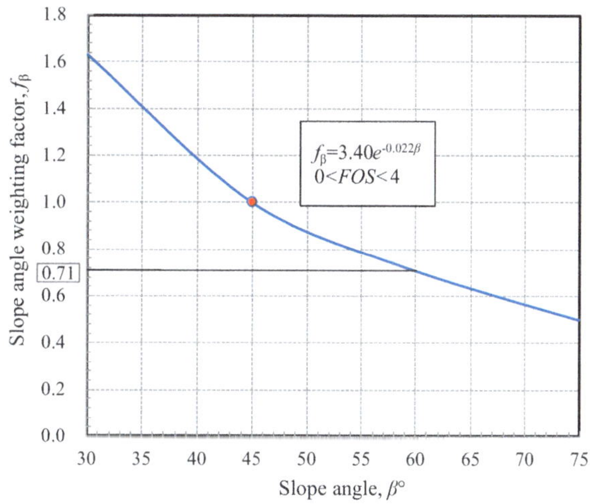

influence of β on the FOS of given slopes based on the PLM for the range β from 30° to 75° while keeping other parameters unchanged. A chart representing the relationship between f_β and β was established based on the data (0 < FOS < 4), which is applicable for the majority of open pit slope cases, as shown in Fig. 12.8.

The use of the proposed stability model to calculate the FOS of a given slope is simple. An example with σ_{ci} = 2.7 MPa, GSI = 50, m_i = 25, γ = 27 kN/m^3, H = 10 m, β = 60°, SR = $\sigma_{ci}/(\gamma H)$ = 10, T/H = 1.0 was used here to demonstrate the use of the proposed model. The evaluation steps are as follows: Step 1: based on the values of SR, GSI and m_i, the value of $FOS_{45°\infty} = 3.20$ is obtained from Fig. 12.6e. Step 2: based on the values of SR, GSI and m_i, the value of $f_D = 0.58$ can be estimated from Fig. 12.7. Step 3: based on the values of SR, GSI, m_i and T/H, the value of $f_T = 1.18$ can be obtained from Fig. 12.3o. Step 4: based on $\beta = 60°$, the value of $f_\beta = 0.71$ can be estimated from Fig. 12.8. Therefore, the FOS of the example slope $FOS_{PLM} = FOS_{45°\infty} \times f_D \times f_T \times f_\beta = 3.20 \times 0.58 \times 1.18 \times 0.71 = 1.55$. Compared with the result of *Slide 6.0* with FOS = 1.52, the discrepancy in this case is only 1.97%.

The prediction performance of the stability model was then tested using 1254 sets of data which cover a wide range of rock properties and slope geometries, as shown in Table 12.4. Compared with the FOS calculated from *Slide 6.0*, the values of FOS_{PLM} estimated from Eq. 12.5 show some discrepancy, as summarized in Fig. 12.9.

The results show that 53.7% of cases have the discrepancy less than ±5%; the absolute average relative error percentage (AAREP) is only 5.6%, and the

Table 12.4 The range of parameters for slope stability analysis

Parameters	Ranges
GSI	10–100
m_i	5–35
SR	0.1–20
T/H	0.5–2.5
β	30°–75°

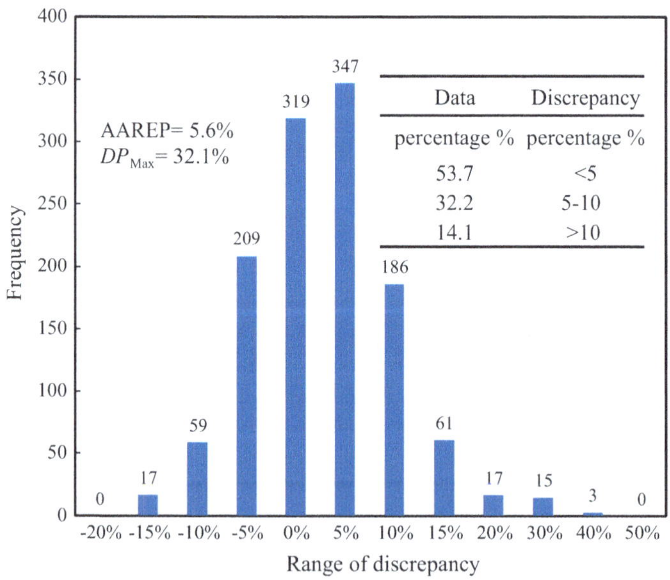

Fig. 12.9 Discrepancy analysis of the proposed slope stability model

maximum discrepancy (DP_{Max}) is 32.1%. On closer observation, it is found that the discrepancy greater than ± 10% appears to be for cases when $\beta = 75°$ and $T = 0.5$ H, which are rare case slope geometries. In these evaluations, D_P is the relative difference between FOS_{PLM} and FOS_{Slide}, as shown in Eq. 12.6. AAREP is a coefficient for assessing the size of relative error and the bias characteristics of the prediction, as shown in Eq. 12.7.

$$D_P = \frac{FOS_{PLM} - FOS_{Slide}}{FOS_{Slide}} \times 100\% \tag{12.6}$$

$$AAREP = \frac{\sum_{i=1}^{N} |D_P|}{N} \tag{12.7}$$

where N is the number of testing data; FOS_{Slide} and FOS_{PLM} refer to the FOS calculated from *Slide 6.0* and Eq. 12.5, respectively.

12.6 Slope Cases Application

The Cadia Hill open pit slope with complex slope geometries and rock mass properties (Franz 2009) was used to further illustrate the use of the proposed stability model. The slope height is H = 500 m and the slope angle is $\beta = 60°$. The slope has three design areas: domain 10, domain 18 and domain 4S, as shown in Fig. 12.10. The rock mass properties and the HB parameters are given in Table 12.5. The blast damage zone thickness T is assumed to be 0.5 H.

In order to calculate the FOS of this slope using the proposed stability model, it is suggested to carry out stability analyses to identify the lower and upper values of FOS. For this case, the values of rock mass properties of domain 10 and domain 4S

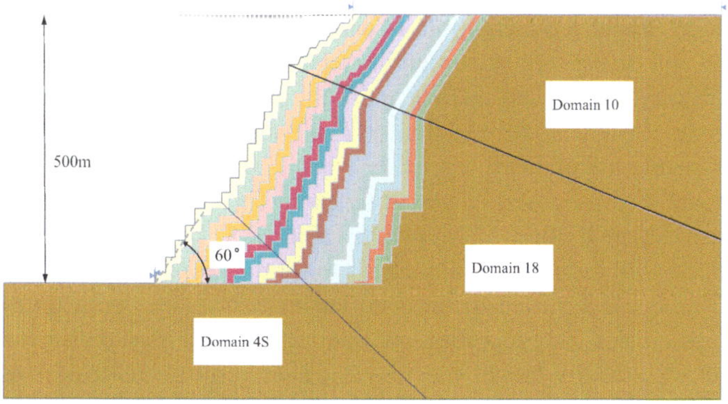

Fig. 12.10 The Cadia Hill open pit slope model

Table 12.5 Rock mass properties of the Cadia Hill open pit slope (Franz 2009)

Properties	Design domain		
	10	18	4S
γ, kN/m^3	27	27	27
σ_{ci}, MPa	90	120	120
GSI	40	46	53
m_i	21	21	21
D	1.0	1.0	1.0

were used to calculate the lower and upper bound values of FOS. The calculation steps are as follows: (1) From Fig. 12.6, $m_i = 20$ and $m_i = 25$ are used to estimate the value of FOS for $m_i = 21$. The values of $FOS_{45°\infty}$ are 2.32 and 2.50 for $m_i = 20$ and $m_i = 25$, respectively. Therefore, the value of $FOS_{45°\infty}$ for $m_i = 21$ is 2.36 using linear interpolation. (2) From Fig. 12.7, SR = 4 and SR = 10 are used to estimate f_D for SR = 6.7, which is 0.51 for this case. (3) From Fig. 12.3, sub-figures *j, k, n, o* are used to calculate the f_T using linear interpolation and the value of f_T is estimated to be 1.40 for this case. (4) From Fig. 12.8, the slope angle weighting factor is found to be $f_\beta = 0.71$ for $\beta = 60°$. Finally, the lower bound FOS is $FOS_{Lower} = FOS_{45°\infty} \times f_D \times f_T \times f_\beta = 2.36 \times 0.51 \times 1.40 \times 0.71 = 1.20$, which is in good agreement with the result of 1.17 calculated from *Slide 6.0*. Following the same procedure, the upper bound FOS can also be calculated and the result is $FOS_{Upper} = 1.76$.

12.7 Conclusions

In this research, the parallel layer model (PLM), which adopts a linearly decreasing D for each layer in the blast damage zone, was used with the limit equilibrium method to explore the effects of blast damage zone thickness T on the stability of open pit slopes under production blast conditions.

Based on comprehensive slope stability analyses, a blast damage zone thickness weighting factor f_T was proposed to quantify the influence of T on the FOS of given slopes. Results show that the selection of T in the slope model plays an important role in the calculation of FOS, especially when T/H is less than 1.0.

A guideline was proposed as a starting point for estimating the f_T of a given slope under different T values, as shown in Table 12.3. The weighting factor f_T was then combined with existing stability charts to form a proposed stability model (see Eq. 12.5) which can be used to estimate the FOS of slopes with different slope geometries and rock mass properties, considering different blast damage zones. The reliability of the stability model was tested against results calculated directly from *Slide 6.0* using 1254 sets of slope cases. The results show that 53.75% of the cases have the discrepancy less than ±5% and the absolute average relative error percentage (AAREP) is only 5.6%. It is also found that the data with a discrepancy of more than ±10% appear to be for cases when $\beta = 75°$ and $T = 0.5\ H$.

Finally, a real open pit slope case with complex slope geometries and rock mass properties was used to illustrate the use of the proposed stability model. The results demonstrated a close agreement between the proposed stability model and the *Slide 6.0* solutions. The relative discrepancy in the FOS is approximately 3.0%. Therefore, the proposed stability model can be adopted as a useful and reliable tool for the preliminary assessment of rock slope stability under different conditions.

References

Bertuzzi R, Douglas K, Mostyn G (2016) Comparison of quantified and chart GSI for four rock masses. Eng Geol 202:24–35

Cai M, Kaiser P, Uno H, Tasaka Y, Minami M (2004) Estimation of rock mass deformation modulus and strength of jointed hard rock masses using the GSI system. Int J Rock Mech Min Sci 41(1):3–19

Deng D, Li L, Wang J, Zhao L (2016) Limit equilibrium method for rock slope stability analysis by using the generalized Hoek–Brown criterion. Int J Rock Mech Min Sci 89:176–184

Franz J (2009) An investigation of combined failure mechanisms in large scale open pit slopes. PhD thesis, The University of New South Wales, Sydney, Australia

Hoek E (2007) Rock mass properties. Chapter 11 in practical rock engineering. https://www.rocscience.com/documents/hoek/corner/12_Rock_mass_properties.pdf. Accessed 20 Apr 2018

Hoek E (2012) Blast damage factor D. Technical note for RocNews. https://www.rocscience.com/documents/pdfs/rocnews/winter2012/Blast-Damage-Factor-D-Hoek.pdf. Accessed 20th Apr 2018

Hoek E, Brown E (1988) The Hoek–Brown failure criterion—a 1988 update. In: Curran JC (ed) Proceedings of the 15th Canadian rock mechanics symposium, pp 31–38

Hoek E, Brown E (1997) Practical estimates of rock mass strength. Int J Rock Mech Min Sci 34 (8):1165–1186

Hoek E, Brown E (2018) The Hoek–Brown failure criterion and GSI—2018 edition. J Rock Mech Geotech Eng. https://doi.org/10.1016/j.jrmge.2018.08.001

Hoek E, Carranza-Torres C, Corkum B (2002) Hoek–Brown failure criterion 2002. In: Hammah R, Bawden W, Curran J, Telesnicki M (eds) Proceedings of NARMS-TAC 2002. Mining Innovation and Technology, Toronto, Ontario

Hoek E, Diederichs M (2006) Empirical estimation of rock mass modulus. Int J Rock Mech Min Sci 43(2):203–215

Hoek E, Karzulovic A (2000) Rock mass properties for surface mines. Slope stability in surface mining, WA Hustrulid, MK McCarter and DJA van Zyl, Eds, Society for Mining, Metallurgical and Exploration (SME), Littleton, CO, pp 59–70

Jiang X, Cui P, Liu C (2016) A chart-based seismic stability analysis method for rock slopes using Hoek–Brown failure criterion. Eng Geol 209:196–208

Li A, Lyamin A, Merifield R (2009) Seismic rock slope stability charts based on limit analysis methods. Comput Geotech 36(1):135–148

Li A, Merifield R, Lyamin A (2008) Stability charts for rock slopes based on the Hoek–Brown failure criterion. Int J Rock Mech Min Sci 45(5):689–700

Li A, Merifield R, Lyamin A (2011) Effect of rock mass disturbance on the stability of rock slopes using the Hoek–Brown failure criterion. Comput Geotech 38(4):546–558

Lupogo K (2017) Characterization of blast damage in rock slopes: An integrated field-numerical modeling approach. PhD thesis, Simon Fraser University, British Columbia, Canada

Marinos V, Carter T (2018) Maintaining geological reality in application of GSI for design of engineering structures in rock. Eng Geol. https://doi.org/10.1016/j.enggeo.2018.03.022

Marinos V, Marinos P, Hoek E (2005) The geological strength index: applications and limitations. Bull Eng Geol Env 64(1):55–65

Melentijevic S, Serrano A, Olalla C, Galindo R (2017) Incorporation of non-associative flow rules into rock slope stability analysis. Int J Rock Mech Min Sci 96:47–57

Michalowski R (2010) Limit analysis and stability charts for 3D slope failures. J Geotech Geoenviron Eng 136(4):583–593

Morelli G (2017) Alternative quantification of the Geological Strength Index chart for jointed rocks. Geotech Geol Eng 35(6):2803–2816

Mohanty B, Chung S (1986) Production blasting and slope stability. In: Proceedings of the international symposium on geotechnical stability in surface mining. CANMET, Calgary, Alberta, Canada, pp 133–140

Peng J, Cai M, Rong G, Yao M, Jiang Q, Zhou C (2017) Determination of confinement and plastic strain dependent post-peak strength of intact rocks. Eng Geol 218:187–196

Qian Z, Li A, Lyamin A, Wang C (2017) Parametric studies of disturbed rock slope stability based on finite element limit analysis method. Comput Geotech 81:155–166

Read S, Richards L (2014) Correlation of direct and indirect tensile tests for use in the Hoek–Brown constant mi. In: Masses-Alejano, Perucho, Olalla and Jiménez (eds) Rock engineering and rock mechanics: structures in and on rock. Taylor and Francis, London, pp 161–166

Sari M (2010) A simple approximation to estimate the Hoek–Brown parameter mi for intact rocks. In: Zhao J, Labiouse V, Dudt JP, Mathier JF (eds) Rock mechanics in civil and environmental engineering. Taylor and Francis, London, pp 169–172

Sari M (2012) An improved method of fitting experimental data to the Hoek–Brown failure criterion. Eng Geol 127:27–35

Shen J, Karakus M, Xu C (2013) Chart-based slope stability assessment using the Generalized Hoek–Brown criterion. Int J Rock Mech Min Sci 64(6):210–219

Shen J, Karakus M (2014a) Simplified method for estimating the Hoek–Brown constant for intact rocks. J Geotech Geoenviron Eng 140(6):971–984

Shen J, Karakus M (2014b) Three-dimensional numerical analysis for rock slope stability using shear strength reduction method. Can Geotech J 51(2):164–172

Sun C, Chai J, Xu Z, Qin Y, Chen X (2016) Stability charts for rock mass slopes based on the Hoek–Brown strength reduction technique. Eng Geol 214:94–106

Tsiambaos G, Saroglou H (2010) Excavatability assessment of rock masses using the Geological Strength Index (GSI). Bull Eng Geol Env 69(1):13–27

Tuckey Z (2012) An integrated field mapping-numerical modelling approach to characterising discontinuity persistence and intact rock bridges in large open pit slopes. Master thesis Simon Fraser University, British Columbia, Canada

Vásárhelyi B, Kovács L, Ákos T (2016) Analysing the modified Hoek-Brown failure criteria using Hungarian granitic rocks. Geomech Geophys Geo-Energy Geo-Resour 2(2):131–136

Wan L, Wei Z, Shen J (2016) Charts for estimating rock mass shear strength parameters. Geomech Eng 10(3):257–267

Wang W, Shen J (2017) Comparison of existing methods and a new tensile strength based model in estimating the Hoek–Brown constant m_i for intact rocks. Eng Geol 224:87–96

Yilmaz M, Ertin A, Er S, Tugrul A (2018) Numerical modelling of steep slopes in open rock quarries. J Geol Soc India 91(2):232–238

Zuo J, Liu H, Li H (2015) A theoretical derivation of the Hoek–Brown failure criterion for rock materials. J Rock Mech Geotech Eng 7(4):361–366

Appendix

The writer is pleased to have been given this opportunity to offer an Editorial to Issue 4, Volume 7, 2015, of our journal, for the first time. The major focus of this Editorial is on the paper, "A theoretical derivation of the Hoek-Brown failure criterion for rock materials" by Jianping Zuo, Huihai Liu and Hongtao Li, published in this Issue. In the following, this paper will be referred to as Zuo et al. (2015). Using this paper as a vehicle, some comments will be offered on this journal's important peer review system. The paper is one of a great many that have appeared in the rock mechanics literature over the past 25–30 years dealing with the Hoek-Brown empirical strength criterion for rocks and rock masses. In the writer's opinion, the present paper is one of the best and most valuable of those papers. Some of these papers have criticised the criterion, others have sought to modify or adapt it in various ways, while still more have used the criterion in developing analytical or numerical solutions to real or imagined rock engineering problems. The great value of the paper by Zuo et al. (2015) is that it considers the basis and derivation of the Hoek-Brown criterion for intact rock using a fundamental, mechanically rigorous, fracture mechanics approach. Interestingly, in a review paper prepared in 2008, the writer (Brown 2008) referred positively to a then recently published paper by Zuo et al. (2008) that is referenced in their present paper. The Hoek-Brown empirical strength criterion was developed in the mid-to-late 1970s during the preparation of the book by Hoek and Brown (1980) in order to fill a perceived need in the tools and techniques then available in rock mechanics and rock engineering. The initial development was the work of Dr. Evert Hoek who, as indicated in the important recent paper by Hoek and Martin (2014), had worked on brittle fracture and Griffith crack theory early in his career. It is not often acknowledged that the original form of what came to be known as the Hoek-Brown criterion given by Eqs. (25) and (26) in Zuo et al. (2015) was chosen to be mathematically similar to the classical Griffith criterion discussed by Zuo et al. (2015) in their Introduction. This mathematical form is reflected by the square root terms in Eqs. (25) and (26).

J. Zuo and J. Shen, *The Hoek-Brown Failure criterion—From theory to application*,
https://doi.org/10.1007/978-981-15-1769-3

The present paper by Zuo et al. (2015) shows how the Hoek-Brown criterion for intact rock material may be derived from first principles using micro-mechanics principles. Importantly, a comparison of Eqs. (24) and (26) shows that the Hoek-Brown parameter, m, may be expressed as $m = \mu\sigma_c/(\beta\sigma_t)$, where σ_c and σ_t are the uniaxial compressive and tensile strengths of the rock, respectively; μ is the coefficient of friction for the pre-existing sliding crack surfaces, and β is a fracture mechanics parameter defined by Zuo et al. (2015).

Brown (2008) showed that, under some reasonable assumptions, $\mu/\beta \approx 1$, so that $m \approx \sigma_c/\sigma_t$. The paper also makes a significant contribution to the understanding of the brittle-ductile transition concept. Importantly, Eq. (28) shows that the confining pressure, at which internal micro-failure processes are completely constrained and the brittle-ductile transition is triggered, is also a function of the parameters μ, β, σ_c and σ_t. It is also shown that the value of the ratio σ_c/σ_t influences the value of the confining pressure at the brittle-ductile transition.

Obviously, the international reputation and standing of a journal such as this depends on the quality of the papers that it publishes. Equally obviously, the quality of those papers reflects primarily the quality of the work carried out by the authors, but it also reflects the quality of the journal's internal review and editorial processes. It is essential that the review process be intellectually rigorous, thorough, honest and constructive, all with a view to ensuring that only high-quality, original papers are accepted for publication. It is one of the responsibilities of reviewers to suggest improvements to the paper where they are able to do so. In this journal, we also ask authors to declare whether or not they have any conflict of interest in submitting the paper for publication. The Editorial Board and Editorial Office depend greatly on the voluntary contributions of the journal's many reviewers. A list of the almost 300 people who acted as reviewers in 2014 was published on the last page of Volume 6, Issue 6.

A feature of the review process is that the identities of the two independent reviewers normally chosen to review each paper submitted for consideration for publication are not revealed to the authors. This is an essential safeguard for the overall integrity of the review process. However, the writer is now about to break this convention of confidentiality in order to make a salient point. The writer acted as one of the reviewers of the paper by Zuo et al. (2015) published in this Issue. Because he was already familiar with the work of Dr. Zuo and his colleagues through their earlier paper (Zuo et al. 2008), because he had a close personal connection with the subject of the paper, and not least because of the overall quality of the paper, the writer found his review task to be straightforward, stimulating and enjoyable. He was pleased to be able to recommend unhesitatingly that the paper be accepted for publication and, as is almost invariably the case even with the very best of papers, he may have suggested that a small number of minor changes be made to the text.

However, when he came to read the "in press" or online version of the paper before preparing this editorial, the writer quickly discovered a small number of errors in the text, one of them not exactly minor, that he had apparently overlooked in his initial review. (Thankfully, the Editorial Office has now corrected these errors in the printed version of the paper.) This was a salutary experience for the writer who immodestly considers himself to be an experienced reviewer and editor having a good eye for detail. It also demonstrates how the quality of published papers depends, in part, on the rigour and thoroughness brought to their important tasks by our many volunteer reviewers. We are most grateful to them and to the authors who offer their contributions for publication in our journal. We all must be eternally vigilant in order to achieve and maintain the highest publication standards.

Milton, QLD, Australia

E. T. Brown
Honorary Editor-in-Chief and Guest Editor of Journal of Rock Mechanics and Geotechnical Engineering
Emeritus Professor, Golder Associates Pty. Ltd.
tbrown@golder.com.au

References

Brown ET (2008) Estimating the mechanical properties of rock masses. In: Potvin Y, Carter J, Dyskin A, Jeffrey R (eds.) Proceedings of the 1st Southern Hemisphere international rock mechanics symposium (Vol. 1), Perth, Australian Centre for Geomechanics. Perth, Australia, pp 3–22

Hoek E, Brown ET (1980) Underground excavations in rock, Institution of Mining and Metallurgy. London, UK

Hoek E, Martin CD (2014) Fracture initiation and propagation in intact rock e a review. J Rock Mech Geotech Eng 6(4):287–300

Zuo JP, Li HT, Xie HP, Ju Y, Peng SP (2008) A nonlinear strength criterion for rocklike materials based on fracture mechanics. Int J Rock Mech Min Sci 45(4):594–9

Zuo JP, Liu HH, Li HT (2015) A theoretical derivation of the Hoek-Brown failure criterion for rock materials. J Rock Mech Geotech Eng 7(4):361–6

Note

The referred papers of Dr. Zuo (Zuo et al. 2008 and 2015) mentioned above have been given thorough demonstrations in Chaps. 2–4, respectively.